# BLAZE

# BLAZE

## NICHOLAS FAITH

CHANNEL 4 BOOKS

First published in 1999 by Channel 4 Books,
an imprint of Macmillan Publishers Ltd
25 Eccleston Place, London SW1W 9NF
Basingstoke and Oxford
www.macmillan.co.uk

Associated companies throughout the world

ISBN 0 7522 1739 9

1 3 5 7 9 8 6 4 2

A CIP catalogue record for this book is available from the British Library.

Text design and typesetting by SX Composing DTP, Rayleigh, Essex
Printed and bound in Great Britain by
Mackays of Chatham plc, Chatham, Kent
Plate section design by Anita Ruddell

This book accompanies the television series 'Blaze'
made by Darlow Smithson for Channel 4.
Executive producer: John Smithson
Producer: Greg Lanning

The publisher has made every effort to ensure accuracy in the reporting of the facts
contained in this book.

To those anonymous heroes, the firefighters.

# Contents

# Acknowledgements

I know this sounds a bit hollow, but it's true: books like *Blaze* and my previous books in this series on other types of investigators are very much team efforts. They are produced on extremely tight schedules and largely based on the transcripts of the interviews which form the bulk of the material in the programmes. So to write the book, I need – and get – a great deal of back-up. This I got to a far greater extent than I deserve from my long-suffering editor, Katy Carrington; transcriber extraordinaire Daphne Walsh; my copy editor, the crisp and decisive Hazel Orme; and from the production team on the series. They – Viv Simpson, Kat English, Jonathan Jones, Caroline Hecht, Emma Jessop, Julianna Challenor, Maxine Carlisle, Sam Alexander and Christabel Nsiah-Buadi – all managed to find time out of their overcrowded schedules to help me promptly, efficiently and without complaint. Thank you all – though I must add that the final result is wholly my responsibility.

# Abbreviations and Glossary

accelerant: a fuel (usually a flammable liquid) used to start or increase the intensity or speed of spread of a fire

AEA (UK): Atomic Energy Authority

ambient: surrounding conditions

ATF (USA): Alcohol, Tobacco and Firearms. Full name is Bureau of Alcohol Tobacco and Firearms (BATF), but commonly referred to as ATF. US government bureau responsible for investigation into fire and fire-related offences

BTU: British Thermal Unit, a standard measurement of heat, the amount of heat it takes to increase the temperature of one pound of water one degree at 60 degrees ambient temperature

calorific value: measure of amount of heat generated by the combustion of a particular material

chromatography: chemical procedure that allows the separation of compounds based on differences in their chemical affinities for two materials in different physical states, i.e. gas/liquid, liquid/solid

conduction: process of transferring heat through a material (or between materials) by direct physical contact

convection: process of transferring heat by movement of a fluid; in convective flow, the warm fluid becomes less dense than the surrounding fluid and rises, inducing circulation

fire-resistive: structure made of materials so constructed as to provide a predetermined degree of fire resistance as they are defined in building or fire-prevention codes, which call for resistance for, say, one, two or four hours

firestorm: overwhelming progression of fire through structures caused by a mixture of convection and radiation

fire wall: a solid wall of masonry or other non-combustible material capable of preventing the passage of a fire for a prescribed time

flame-over: the flaming ignition of the hot-gas layer in a developing compartment fire

flame resistant: material or surface that does not maintain or propagate a flame once an outside source of flame has been removed

flammable: see inflammable

flashback: the ignition of a gas or vapour from an ignition source back to a fuel source (often seen with inflammable liquids)

flash-over: the final stage of the process of fire growth; when all combustible fuels within a compartment have ignited the room is said to have undergone flash-over

FPA (UK): Fire Protection Association

HSE (UK): Health and Safety Executive

inflammable: a combustible material that ignites early, burns intensely, or has a rapid rate of flame spread

NFPA (USA): National Fire Protection Association

NIST (USA): National Institute of Standards and Technology

pyromania: uncontrolled psychological impulse to start fires

radiation: transfer of heat by electromagnetic waves

seat of fire: area where the main body of fire is located

spontaneous ignition: chemical or biological process that generates sufficient heat to ignite the reacting materials

volatile: a liquid with a low boiling point, one that is readily evaporated into the state of vapour

(Many of these definitions are taken from *Kirk's Fire Investigation* by John D. de Haan, Prentice Hall, Upper Saddle River, New Jersey)

# Interviewees

Stan Ames, fire consultant
Jose E. Aponte, witness, Du Pont Plaza
Steve Avato, special agent, ATF, Philadelphia
Bob Bell, senior forensic scientist, Forensic Science Service
Mike Brouchard, chief, arson and explosives program, ATF
Bob Buckley, ex-fire investigator, Philadelphia fire marshal's office
David Canter, professor of psychology, Liverpool University
Marvin Casey, Bakersfield fire investigator
Nelson Chamfrau, in charge of emergencies, New York Port Authority
Dr Lau Chau-ming, senior chemist, Forensic Science Division
Bobbie K. K. Cheung, investigator, Hong Kong Government Forensic
    Service
Vivian Chien, survivor, Garley Building fire
Kathleen Collins, survivor, World Trade Center explosion
Ed Comeau, chief fire investigator, NFPA
David Cowen, writer
Wayne Dammert, survivor, Beverly Hills Supper Club fire
Ralph Dinsman, battalion chief, Clark County Fire Department, Las Vegas
Dr Joseph Dreyfus, survivor, Coconut Grove fire
Carl Duncan, independent fire analyst
Rich Edwards, investigator, Los Angeles County sheriff's department,
    arson explosives detail
Mike Ellis, senior pilot, Government Flying Service
William Feinberg, professor of sociology, University of Cincinnati
Jim Ford, assistant chief, Rural/Metro Fire Department
Ed Galea, fire-safety engineering group, University of Greenwich,
    London

Ron George, latent fingerprint examiner, Los Angeles County Sheriff's Department

Paul Godier, head of safety and environmental development, London Transport

Bob Graham, MBE, ex-chief fire investigator, Greater Manchester Fire Brigade

George Gravey, firefighter

Casey Grant, assistant vice-president, NFPA

David Halliday, senior forensic scientist, Forensic Science Service, UK

Steve Hanson (USA), deputy fire chief, Clark County Fire Department

Steve Hanson, police officer, King's Cross, London

Norris Johnson, emeritus professor of sociology, University of Cincinnati

Dr Ian Jones, CFX chief technologist, Atomic Energy Authority Technology

John P. Keating, professor of social psychology, University of Wisconsin

Tom Klem, fire protection engineer

Dr Lau, site investigator, Hong Kong Government Forensic Service

Lieutenant Tommy Lawson, Philadelphia fire marshal's department

Sergeant Glynn Leesing, West Yorkshire Fire Brigade

Dr Kenny Leung, dental surgeon

Glen Lucero, arson investigator, Los Angeles City Fire Department

Bill Malhotra, former head of buildings and structures division, Fire Research Station, UK

John Malooly, special agent, ATF

John McCool, survivor, Philadelphia fire

Father Heber McMahon, survivor, Stardust disco fire, Dublin

Tony Marra, survivor, Coconut Grove fire

Mike Matassa, special agent, ATF, Los Angeles arson task force

Rebecca and Rachel Mizioch, saved by sprinklers

Keith Moodie, head of fire/explosives group, Health and Safety laboratory, UK

Penny Morgan, senior fire consultant, Centre for Fire Protection Systems at the Fire Research Station

Bud Nelson, NIST, Puerto Rico

Harold Nelson, research fire protection engineer

Sue Osbourne, firefighter

Lieutenant Renney Pelszynski, assistant fire marshal, Philadelphia Fire Department

Chris Porreca, manager, arson and explosives program, ATF
Seamus Quinn, detective sergeant, Dublin Garda
Alan Reiss, manager, World Trade Center, New York
Dick Reisenberg, fire chief in 1977, Beverly Hills Supper Club fire
Ian Rogers, King's Cross Station manager
Malcolm Saunders, deputy fire chief, West Yorkshire Fire Brigade
Richard Scheidt, firefighter
Rob Seaman, firefighter
Peter Shilton, divisional officer, projects manager, Avon Fire Brigade
Bob Silverman, survivor, Du Pont Plaza fire
Suzanne Simcox, mathematician, Atomic Energy Authority Technology
Dr Jonathon Sime, environmental psychologist
Joe Swartz, fire investigator, NFPA
Neil Townsend, divisional officer, London Fire Brigade
Andy Vita, associate director, ATF
Dian Williams, president, Center for Arson Research, Philadelphia
Steve Wood, survivor, Woolworths fire
George Yaeger, battalion chief, Philadelphia Fire Brigade

# Introduction

So near the fire as we could for smoke; and all over the Thames, with one's faces in the wind, you were almost burned with a shower of the drops. When we could endure no more upon the water, we to a little ale house on the Bankside and there staid till it was dark almost, and saw the fire grow; and as it grew darker, appeared more and more; and in corners and upon steeples, and between churches and houses, as far as we could see up the hill of the City, in a most horrid, malicious, bloody flame.

Samuel Pepys' description of the Great Fire of London in 1666

Fire is the most fearsome phenomenon encountered by those living even in the most modern, most supposedly civilized of surroundings and is therefore the most shocking of all the types of disaster discussed in my previous books, on the sea, our roads, and in the air. In all these cases we are more or less aware of possible dangers. But when we are at home, in an office, a department store, a casino, a hotel, a nightclub we think we're safe: we have no apprehension of danger, which makes a disaster seem much worse.

Of course, like *Black Box*, *Crash* and *Mayday* this book is not about the blazes but about what happens once a fire has been put out, the survivors rescued and the victims carried away. It's about how investigators try to carry out the impossible task of ensuring that such disasters never happen again.

The smoke and hot gases generated by fire are deadly enemies. If the fumes meet an obstruction like a ceiling or roof, they spread out laterally until they reach a wall. Smoke, in particular, will discover any hidden cavities or voids and appear in places far removed from the source of the fire. Even an apparently sturdy material, like steel, can help to spread a fire, because the heat is conducted through the metal, or it may lose its strength when heated and then distort and collapse. 'Fire spread' occurs when the hot products of combustion move through a building. Most serious is when the smoke contains large quantities of unburnt gases. When mixed with air these may ignite and explode, leading to 'flash-over' – when all the combustible materials in a room catch light spontaneously in a racing ball of fire.

Fires, of course, have a multitude of causes – although Canadian statistics suggest that out of ten, seven are due to the hand of man, either accidental or deliberate, one to products, processes or materials, and the remaining two to defects in buildings. Heating, cooking, smoking, lighting fires or even matches, together with the electrically generated fires so common in vehicles dominate the statistics. People die in fires for two reasons: they are either burned or asphyxiated, because fire consumes all the available oxygen. In temperatures of above 3000°C, which are reached in many fires, people die within a few minutes. In addition, smoke may contain toxic products or irritants that attack the mucous membrane of the respiratory tract. Dozens of toxic products may be released by fire, ranging from ammonia to sulphur dioxide and, of course, the searing flames themselves.

The problems encountered by regulators, firefighters and investigators have been exacerbated by the swift advances over the past half-century in building techniques, building materials and the size and complexity of new buildings. The devising, enacting and enforcing of new regulations for the environment has not kept pace with progress.

One problem, perhaps, is that we are not sufficiently afraid of fire. As an American report put it in 1973:* 'Crime may stalk the city streets, but certainly not fire.' The authors of the report attribute this smugness partly to the fact that there have not been any major urban conflagrations for more than half a century and partly to the 'newness of so many buildings [which] conveys the feeling that they are invulnerable to attack by fire'. Further, they say, 'Americans tend to take for granted that those who design their products, in this case buildings, always do so with adequate attention to their safety.' It is a statement that might apply to everyone in the civilized world, and it is not a valid assumption.

In the front line are the firefighters, the Poor Bloody Infantry in the

---

*Report of the National Commission on Fire Prevention & Control, Library of Congress, 1973.

eternal war against fire, their role glimpsed but not fully described throughout this book. Their job has always been hazardous. At one point a mean-minded authority tried to define a firefighter's job as being 'semi-skilled'. The riposte from the firefighter's barrister came in the form of a quotation from that most famous of Victorian firefighters, Sir Eyre Massey-Shaw:*

> A fireman, to be successful, must enter buildings. He must get in below, above, on every side; from opposite houses, over back walls, over side walls, through panels of doors, through windows, through skylights, through holes cut by himself in gates, walls and the roof. He must know how to reach the attic from the basement by ladders placed on half-burnt stairs, and the basement from the attic by a rope made fast on a chimney. His whole success depends on his getting in and remaining there, and he must always carry his appliances with him, as without them he is of no use.

Today firefighters are far better protected than they were in the past, but their job remains lonely and heroic and not unlike that described by Sir Eyre a century and a half ago. Not surprisingly they form strong personal bonds – as we shall see in Chapter 8, with the death of Fleur Lombard. I appreciated the strength of just such a bond in one firefighter's description of the death of another who had gone back to rescue his wife in the fire in the Beverly Hills Supper Club, discussed in Chapters 12 and 13 (with tragic irony he had not been aware that she was already safe). 'He was a firefighter,' said his colleague, 'and I don't mean to say that his death is any more important than the others because it is not, but to me it *was*, because he was one of my firefighting brothers.'

We are, aware, however vaguely, of the role played by firefighters, and we mourn the deaths of the victims of fire, but we forget the agonies of the survivors, which may end in breakdown or even suicide of those who consider themselves, or are considered – usually wrongly – responsible. For even those who have played a heroic role, as Father Heber McMahon in the fire in the Stardust disco, suffer in the immediate aftermath of the blaze:

*Immortalized as the Captain Shaw serenaded by the fairies in Gilbert and Sullivan's *Iolanthe*.

'I was living in a house on my own and I opened the hall door, I came into the kitchen, turned on the kettle then went back into the sitting room. I always remember the chair was in front of a dead fire. It was now about seven in the morning and I was just sitting staring into the empty fireplace and I was kind of talking to myself, "That's awful, unbelievable, what happened last night," and another part of me was saying, like "There's no way that you, Heber McMahon, could have done what you think you did last night, you know, you're an ordinary guy and here you think you were pulling people out of fires." Eventually I was going slightly crazy, I think, wondering was this just one huge nightmare, and I always remember the simple thing, I was frozen with the cold anyhow because I'd been out in the open a lot of the night and the shock, I suppose, but I still had this anorak on and I just lifted up the sleeve and I smelt the smoke and I said, "It happened, it happened, you were there."

'I then called next door, we had sisters in the parish, nuns, they were all medical-trained, so I waited for about an hour. It was just before eight and I rang the doorbell, just because I needed somebody. I was bursting, like, at this stage with all the emotions, and one of the sisters opened the door and I said, "Can I come in for a moment?" She was looking at me rather strange, eight o'clock in the morning, and I'm not an early bird ever, so I said, "Did you hear anything last night?" She said, "Why?" I said, "The Stardust went on fire last night, there's an awful lot of young people dead." She said, "No", and I said, "Yeah, you must have heard something." At this stage I began to see this woman, who was a surgeon, incidentally, a tough cookie, I could see her looking at me and thinking, McMahon is losing it, he's obviously hallucinating, he must have been on the bottle last night or something, so just then I said, "Turn on the news", so we turned it on and, of course, it was the top item. I was kind of relieved.

'It would be great in a sense,' he continued, 'if an awful thing happens, if it can be allowed to happen and be finished with, you know, over and done with, and that people can get their lives back together again, but life isn't like that and something as tragic as that affects everybody. It's a tight-knit community – Artane, and everybody knew one of those nearly fifty young people who died.'

The funeral, held on a day of national mourning, helped: it was, he said, 'a very good way of acknowledging that something dreadful had happened, that it's not something that the people of the community themselves were expected to cope with on their own. There is that sense of "You are not alone." That was a moment of hope for me and I think for a lot of other people too. They said, "Look, life maybe can go on." '

# I
# THE INVESTIGATORS

# 1
# The Investigators

It is now over thirty years since, as a forensic scientist, I attended my first fire scene. The site was a woollen mill in the West Riding. It was dull, cold and raining. I stood on the ground floor of the ruined building, surrounded by carbonized debris and looked up at the sky and the blackened beams from which water dripped to form steaming puddles at my feet. I recollect the question I asked myself, 'How can anyone extract useful information from this situation?'

Stuart S. Kind, President of the International Association of Forensic Sciences

The experience was enough to turn Dr Kind off fire investigation for life, but he did define the requirements for investigators: 'Substantial experience of fire-scene investigation, a flair to make useful subjective judgements and the ability to set up useful lines of inquiry and informative scientific experiments to negate, or add confirmatory evidence to, a hypothesis of cause. Thus the procedures of the fire examiner are exactly similar to those of the detective investigating a crime or the aeronautical engineer investigating a flying accident.' Kind was being quoted in the introduction to a manual on the subject by H.T. Yallop,* who adds a number of other qualities, notably a knowledge of the methodology of fire investigation, a relevant academic qualification – most usefully, in his opinion, a degree in chemistry – and back-up from adequate laboratory facilities.

Other investigators give even longer and more daunting lists; reading them makes you wonder if anyone except a superhero could be worthy of inclusion, and there are few, if any, female investigators. For Chris Porreca of the ATF†, the investigator 'needs to be thorough, methodical, able to

*Fire Investigation, Alan Clift, Solihull, 1984.
†The Bureau of Alcohol, Tobacco and Firearms, commonly shortened to ATF.

think in the abstract and to walk into something that is in essence a black hole. He'll look into it and everything is charred and burnt and he needs to be able to abstract in his mind, take a look at it and start to put the pieces together. He needs to be able to say that this was this and that was that and to just work through all the different scenarios that could have happened here but not develop tunnel vision. Not one of us walks through a doorway and in the first ten seconds says, "That happened right there", because you can't do that. You need to be able to take all the information, build the scenarios and then make the final determination. If you walk in and get tunnel vision you're not allowing yourself the opportunity to do everything you need to do, so it's being open-minded and being able to picture things abstractly.'

For Mike Brouchard, chief of the ATF's arson and explosives program, 'A good fire investigator has to be thorough, patient, curious, inquisitive, and they can't take no for an answer, they have to be very dogged on an investigation and be willing to spend about a year on some cases to determine if they can bring a prosecution to court.' By contrast, the definition given by John Malooly of the ATF's special response team is one that I would associate with those of a good journalist: 'Perhaps the most important quality would be curiosity. If a fire investigator is not curious he will not make a good fire investigator. You see many things when you're doing your fire-scene examination and you have to be innately curious because you want to try to establish what caused these various things.'

Ex-fire investigator Bob Buckley is more down-to-earth: 'A good fire investigator has to combine the skills needed to investigate a fire, which would include experience fighting fires, training in interviewing techniques, report writing, with intuition – maybe you should suspect everything and not assume anything early on. By being in the fire service you see first-hand what fire does, how it moves, and people will tell you a lot of things. You can relate to them when they're telling you what they saw, how things reacted. You have to know when people say, "This exploded", that maybe it wasn't an explosion, it was a window blowing out. We can relate to that because we've been in fires where we've seen this happen. So the first-hand experience of actual firefighting is, I think, the biggest thing a fire investigator needs. It takes a long time to be a good firefighter and it takes a long time to be a good investigator.'

Perhaps experience and intuition are the prime qualities required for any investigative profession, but in addition, and this emerged time and again in interviewees' statements, there was rigorous professional discipline, the refusal to take anything for granted or to take shortcuts, or to assume that the first, or the most inviting, answer to the question thrown up by a fire was necessarily the right one. In the words of Andy Vita, associate director of the ATF, which could have been spoken by any of his colleagues on either side of the Atlantic, 'You can't have a preconceived notion of what happened. You have to let the evidence direct you to the solution.'

The scientific investigation of fires has developed only over the past twenty or thirty years. Although in some form or another fire brigades have been around for over two thousand years, since Roman times, the art and science of the investigator – and even the philosophy of fire prevention – is a more recent phenomenon. Suitably enough its pioneer was James Braidwood, one of the greatest firefighters. He had already revolutionized firefighting by insisting that his men try to get to the heart of a blaze rather than standing well away from a burning building and pouring water in through the windows.

In 1832 Braidwood had been appointed to command the brigade formed in London from the amalgamation of the many units financed by insurance companies. He immediately started to inspect the inside of buildings he felt might be at risk in order to suggest preventive measures and to ensure that he could work out how to fight a fire if it did break out. (Unfortunately he did not have time to pursue his ideas with the old Houses of Parliament – though he succeeded in saving Westminster Hall – when they caught fire two years later.)

But Braidwood's pioneering work was on fire prevention, which grew in importance with the increase in size of buildings and use of more modern – and usually more fire-resistant – materials. The lag in fire detection, as opposed to firefighting itself, can be shown by the fact that even in the United States the National Commission on Fire Prevention and Control was established only in 1971 and its report, published by the Library of Congress two years later, was a horrific document, revealing wide gaps in every aspect of preventing, fighting and investigating fires.

Today, however, most major cities – in Britain as well as in the USA – employ specialist fire-investigation officers, and specialist national organizations investigate fires and try to improve the fire-resistant qualities of buildings and their contents. In Britain, these include the Health and Safety

Executive (HSE) and the Forensic Science Service. In the USA the national investigatory teams are under the aegis of the ATF, but, in theory, have the power to assist only where 'interstate commerce' might be involved, which includes most major commercial or manufacturing premises. However, they're often called in, as Mike Brouchard puts it, 'to assist state and local investigators in investigating arsons that involve crimes such as arson rings or conspiracies'. In 1978 they even formed a national response team to ensure that specialists would be available in the crucial hours immediately after a fire. They arrive with some of the many tools listed by Robert E. Carter in his book *Arson Investigation*.* These range from the simple and obvious – saws, lights, tape-recorders, pens and pencils – to combustible-gas detectors and portable gas chromatographs.

The investigators' task is not made easier by the variety of causes of fires. In *The Thin Red Line*† Stephen Barlay gives a selection:

Faulty or misused electrical installations and appliances (unprofessional connections, arcing, faulty earthing, frayed wiring, overloading of power points, etc.); heating appliances; faulty, overheating and sparking machinery; naked lights (workers with oxy-acetylene cutting equipment, welders, and maintenance staff with blow lamps playing a predominant role); and deliberate rubbish incineration. In homes, heating and electrical appliances are major sources, while at some places of work spontaneous combustion (in hay, for instance) can be a serious problem.

Moreover, fire can spread through convection, by conduction to other flammable materials nearby and by radiation, when strong heat radiates across an open space igniting materials with which it has no apparent direct physical connection.

Bob Buckley sets out the methodology required to find a path through this thicket of false clues in the classic sequence: 'The first priority,' he says, 'is to decide where the fire originated, and that's done by interviewing witnesses who were in the area and had first-hand knowledge of the initiation of the fire. We inspect burn patterns in the building, working from the least damaged to the most damaged area, and following the patterns to a specific area at first.

*Collier Macmillan, London, 1978.
†Hutchinson Benham, London, 1976.

Then you can narrow it down to an actual origin. To do that we move to the investigation of where the fire started. Once you've established where it started then you stand some chance of being able to find out how. Clearly the fire will have spread from where it began until it involved a whole building but you have to go back from the point to where the fire has developed to where it began and then you can probably find out how it started.'

As Chris Porreca, manager, arson and explosives program, ATF, puts it, the investigator 'almost works backwards. He takes a look at the last actions that occurred which, in the instance of a fire, would be the arrival of the Fire Department and what they did. You have to take a look at the overall scene and you have to assess the damage to the area and to the structure. You need to ask the Fire Department what it looked like upon their arrival, how did they fight the fire, how did they ventilate it, and what were their primary concerns when they arrived there. Was their primary concern a rescue? Were people trapped in the building? Or was it protecting other structures? Or to get into the building and attempt to extinguish the fire? Each fire has a different set of priorities that the Fire Department has to assess when they arrive, and their actions in the first few minutes of the fire could drastically alter the fire scene.

'Then you would start to work your way back into the scene. Who were the last people at the building? What did they do right before they left it? You begin to take a look at the area of least damage. In a large structure you might get an end of a building or a section of building that has no damage or just light smoke damage but you still examine that area of the building and you work your way to the areas of more damage until you get to the most damaged area. You now have to look at that and determine if that is the area the fire had started in or if there is another reason why there's that much damage there – which depends on what we call a fuel load, which would be furnishings, materials, what is in that building. You might get an area of a building that is damaged significantly because of what was stored there. It might not necessarily be where the fire started but because of the fuel there you get more damage. Our investigators are trained to take all of these factors into account before they render their opinion as to what the area of origin of the fire was and what ultimately caused the fire.'

Next comes close study of the pattern of burns in the fire. 'Burn-pattern analysis,' says Buckley, 'is used a lot in fire-investigative techniques to determine the origin of the fire and the path it travelled. Most times when a fire starts the heaviest area of burning is where the fire originates and then the heat moves up and out from that location and ignites other combustibles. The available fuel load in the building will determine how fast and how severe the fire is. Generally there's a quick response to the fire so there is a lot of unburned and incompletely burned material. Then we look for the lowest point of burning, the heaviest point of damage and work out how the fire moved from that area. Sometimes one side of a cabinet or a table is more heavily burned than the other side, which shows a direction of travel. Ceiling joists and beams can be used to analyse which way the fire moved.'

Once the investigators have identified the rough area where the fire started they begin the laborious process of sifting through the debris in that area: 'Sorting through the debris in a room to find out what was in it and to determine a possible ignition source is a lot like doing an archaeology dig,' says Buckley. 'You sift through it a layer at a time trying to identify what's on top and removing the top layer and working your way down, trying to identify each item. Some you'll be able to identify, ceiling tiles, a metal grid from a ceiling, lightbulbs, door hinges, and you mark their location just like you would in a dig to see where it was when you found it and what it means. Does the hinge mean the door was open? Was the door handle found inside a room or by the frame? This can tell you a lot and it can tell you about the materials that were in there. You might not find the full remains of, say, a vacuum cleaner but you might find a motor so you know that we had some motorized equipment there. It's just going through layer by layer identify-ing what you find, where you find it and should it have been there?'

After documentation and analysis, the investigators bring out the mass of charred fragments. After a fire in Puerto Rico (see Chapter 2) Andy Vita says that he and his colleagues 'set up "sifting screens", a series of screens with different-grade mesh that we can put the debris through. If there is evidence of a bombing pieces of the bomb and the initiating device are normally found in that debris somewhere. You'll find pieces of wire, per-haps pieces of a clock, you may find wrapping from dynamite, bits of pipe and various components of explosive devices. It's amazing how much stuff remains when you think a fire consumes all the evidence. For a trained investigator a lot of things can be found that will be indicators as to the cause of a fire. We keep track of everything recovered, and develop a grid in that area so we can determine where exactly each wheelbarrow of debris came from so that we can reconstruct [relative] distances from the [fire] site of the scene.' It was only through such laborious work that at Puerto Rico – and the same is true of many other such investigations – they were able to prove that there hadn't been an explosion.

After all their painstaking work, investigators must also be able to contemplate the possibility of failure. 'Not all fires are determinable,' says Chris Porreca. 'There are certain fires that when you go and examine them you might get to where the fire started but the actual cause of the fire isn't always determinable. The challenge is to perform the best possible examination and then to determine if it was a humanly induced act, an incendiary fire, or an accidental fire?

Grim, and sometimes futile, though their work can be, it can also be immensely satisfying. 'The satisfaction,' says Porreca, 'is in putting the puzzle together. A lot of us look at it as a puzzle. You walk in and everything is burned and there's a lot of things you can't immediately identify. You slowly examine everything in the room and put the parts back together again. The satisfaction is being able to come up with a determination.'

Bob Buckley says, 'I look at it as like it's a puzzle to solve, and a lot of times that's exactly what it is. You take material that was thrown out of the building by firefighters, you bring it back in, you reconstruct it and you're rebuilding the room the best you can. A lot of times when you do that the point of origin stands out. It's like you just put in the last piece of the puzzle and there it is. The satisfaction is that you were able to come up with the cause, and you could use it maybe in a fire-prevention programme to make other people aware that, hey, we're getting a lot of problems with kerosene or electricity, and you may save some lives down the road.'

But, Porreca, as for many of his fellows, the biggest satisfaction lies in cases of arson 'in which someone has been injured and we've worked the investigation to catch the individual. Then when we get the actual prosecution we know that we've done some good, that someone is locked up who caused a death or injury to someone else.'

John Malooly gets 'a lot of personal satisfaction out of it. When we arrive on a scene there's a lot of chaos, there's a lot of damage been done, maybe injuries, possibly loss of life. Nobody really knows what happened or how and it's an intellectual challenge to try to reconstruct the scene, the room, the building and the series of events that led to that small initial fire which resulted in maybe a high-rise building burning. There's a lot of satisfaction in being able to carefully reconstruct by means of the physical examination and interviews exactly what series of events caused the fire to begin.'

He goes on: 'The longer our fire is burning the less evidence we have left. A burglar commits a crime, he leaves evidence when he departs, and it's pretty much a stable event. If he's left fingerprints they're going to be there. The longer that the fire burns, the less evidence we have left to look at and the harder our job is as time progresses. It gives us a lot of satisfaction to be able to reconstruct that event and determine that this was an arson fire and then do the investigation to catch the arsonist, bring him to justice and put him in jail where he belongs.'

Steve Avato, special agent with the ATF, was the most enthusiastic: 'I love this job,' he says. 'I think this is one of the best jobs in the world, it's hard to believe because we often go to scenes that involve destruction and death but there is a lot of satisfaction in figuring out what happened here and trying to prevent it from happening again. I like to look at it from the standpoint that firefighters have risked their lives to put this fire out and now I want to contribute something to that cause, I want to give them back something and I like to try to figure out what happened so we can go back to the firefighters and tell them, then use it as part of an education programme for the public to prevent that type of fire from ever occurring again.'

# 2
# The Arrival of Science

What really happened at the Du Pont Plaza was the emergence of a scientifically driven analytical method using fundamental principles of chemistry and physics applied to the dynamics of fire growth and this was a significant watermark in fire protection engineering where not only was the classic approach to the investigation and analysis of this fire achieved but so too was the application of the fundamental engineering principles of fire growth dynamics.

Tom Klem, fire protection engineer

One of the major hotels in the popular Condado Beach section of San Juan, the capital of Puerto Rico, was the twenty-two storey Du Pont Plaza Hotel. The winter holidays mark the peak of the tourist season in Puerto Rico, and at the end of 1985 the hotel was fully booked. On New Year's Day 1986 a fire devastated the hotel, leaving ninety-seven people dead and property losses of millions of dollars. The only lucky factor in the tragedy was that the fire broke out in mid-afternoon: the hotel had registered over a thousand guests, and although several were on the beach many diehard gamblers were in the casino. In the words of Andy Vita, an associate director of ATF, 'Had this fire occurred late at night when all the guests were asleep, the death toll would have been far, far greater.' In the end the fire, and the investigations that followed, marked a major step forward in the scientific treatment of such disasters and gave rise to a more profound understanding of all the elements that make up a major conflagration.

The fire developed in a stack of furniture in one of the hotel's two ballrooms. A folding, accordion-like wall divided the north and south ballrooms, and the fire started in the south ballroom area. It would probably not have ended in such tragedy had the building not been altered, without permission. The damages did not affect the ballroom where the fire was and the casino that held most of the victims. But: 'As originally designed,' says Bud Nelson of the National Institute of Standards and Technology (NIST), 'the foyer was open to the ocean and vented at the top. At some time a glass wall was erected to separate the foyer from the ocean and glass was used to fill in the vents at the top. Had the wall not been there at the time of the fire when the windows broke out in the south ballroom it would most likely have vented itself out towards the ocean and not have intruded into the casino. With the enclosure the materials held the smoke, the flame and the hot gases inside the foyer, allowing them to build up heat and penetrate into both the hotel lobby and into the casino. Had the enclosure been in the original design they never would have used those glass walls. They would possibly have been masonry walls to separate the south ballroom and the casino. Had they been windows, they would have been built of either wire glass or some other materials that can resist fire but they weren't. They were built as if they were outside windows so they broke quite readily under fire conditions.'

As a result the fire could travel from the ballroom to the casino through a foyer that encouraged its spread because it had a wooden ceiling. The fire developed in the ballroom. It was spotted but spread fast because, as there was no legal requirement at the time, there were no sprinklers and no fire-resistant doors between the foyer and the casino. Had the foyer not been enclosed, the ocean breezes would have dispersed the flames so that less spread to the casino. As Vita explains, 'The natural air currents, especially those from the ocean, helped move the fire and gases out along past the elevators then up into the second storey. The fact that the lobby area, which is on the second floor, is open at three sides helped to move the fire and influenced its progression through the building. As the fire went up the stairs that led from the first to the second floor it passed through that open-air area and vented out straight through the casino and the windows.

'As the fire developed,' Vita continues, 'people moved through the casino to exit out into the lobby area and hopefully then down the spiral staircase to the pool level. We found bodies on this corner of the building both at the wall by the windows as well as by the door that led into the hallway. The door itself was a wooden one that required two functions to open it. A latch had to be moved downward and pulled towards the opener. Once the smoke moved into that room it was dark and very difficult for people to see. They probably felt their way around to this side and there was probably panic. As they tried to get down to the wall and out to the doorway the pushing at the door prevented the first person from pulling that door back towards them. With all the people trying to exit that room it was just impossible to do that. A number of people realizing this started to move back to the casino. By the time they were able to open that door the flame front had moved through there and vented out through the windows.' It did not help that at least one door was locked. After the fire Bob Silverman, a survivor, found that this 'was a normal thing for casinos to do if there's some sort of panic'.

'In the end,' says Vita, 'fifty-five bodies were recovered from the area round the west door and another twenty-five in the corner of the room.'

The conditions within the casino were truly terrible. In the words of a regular guest, Jose E. Aponte, 'The smoke was starting to come in through the air-conditioning ventilators. At first it was coming slow, like when you smoke a cigarette, but then something happened downstairs at the ballroom that the smoke started coming into the casino faster and faster and faster and all of a sudden almost everything inside was dark, almost black. I passed out because I got choked with the smoke. It was poison, it was hot, it was like acid, it was like when you burn plastic and other material. It gets into your throat, into your stomach, into your lungs and you can't breathe, you choke and the more you try to breathe the more you choke.' Like a number of other survivors Aponte jumped from the casino, breaking his legs in the process. Even then help was not immediately forthcoming. His rescuers carried him to the Dutch Inn Hotel across the street. When they got him there, he claims that one of the employees said, ' "Look, don't bring that person in here, he is bleeding, his blood is going to stain the rugs," so they decided to take me to the Pizza Hut.'

The Puerto Rican authorities asked immediately for help from the ATF, which put together one of the biggest emergency teams it had ever assembled to investigate the fire as well as individual analysts from the US FPA and fire-spread specialists from the Centre for Fire Research. First on the scene was Andy Vita.

Although most of the rescue work had been completed by the time he got there, Vita found that 'There were a few smouldering areas where you could see the smoke coming up. The majority of the people had stopped their work for the day because it was into the evening when we arrived. . . . You could hear the wind blowing and the waves behind the building coming off the ocean and it was rather eerie to sit and look at that building and realize that there was so much destruction in it.

'The local police, the FBI and other investigators had already moved on to the scene so Vita emphasized to them that 'We needed first to identify the source and the cause of the fire [because] there was always the possibility that it could have been accidental.' Because, as Vita says, 'We were the new kids on the block when we arrived and everybody else had their own ideas and own agenda and own motives for conducting their investigation, we had to educate them on what assets we brought and what value we could be to that investigation. Within a very short time, I think, we were able to convince the people there of our technical expertise as well as our investigative abilities, and we started to work together more as a team.'

Their first concern was whether the damage had been caused by an explosion. 'There was a great deal of hostility on the island at the time and it was thought that it could have been a terrorist act. We knew that there were some union negotiations that were ongoing with the Teamsters Union in the hotel, we knew that there had been a series of about three fires that had occurred within the ten days preceding this major fire. They appeared to be nuisance fires but they were intentionally set fires. They had not yet determined who was responsible for them but there was a thought that that was a part of this whole thing with trying to influence the owners of the building to acquiesce to the union's requests.'

Some 250 workers had contracts that came to an end on 1 January. At about 2 p.m. on 31 December, union officials met management in the north ballroom. At the meeting, 'a couple of fights broke out in the ballroom where it was being held. There was some thought later that that was a diversion created to draw attention away from the area in which the fire was set. After the fight the people were removed from that area and there was nobody in the back of the ballroom and that's when we believe that the fire was set. The fire probably began within a half-hour after the ballroom was evacuated.'

'Within the first day,' says Vita, 'we were able to identify that there seemed to be an area in the ballroom which was on the first floor of the building towards the back,' where the investigators found 'debris from various pieces of furniture that had burned. We found that just prior to the fire the hotel had received stacks of new furniture [which] was piled up in that area.'

After interviewing the firefighters to find out how the fire had spread, 'we began our investigation,' says Vita, 'by going to the area of least burn and highest burn, to the area of greatest heat and greatest burn damage. When we looked at the video footage we could see the fire had moved from the ballroom area in the lower level of the building and appeared to move through from left to right at the back of the building, it moved through the second floor and went out of an area that was adjacent to the casino. So we started walking the fire back from the exit by the casino doors, back through the building, down the stairs and into the ballroom, where we could see that in the far corner of the ballroom there was extensive damage both to what appeared to be a stack of furniture and the ceiling material above it, which showed deflection and damage to the large steel structural members. This indicated there was intense heat and a long exposure to that heat in that area, so that's where we kind of focused our investigation. We looked at that area as being the area of origin of that fire. We set up a series of variously graded screens where we put the debris. Then we sifted the material down through these screens. The larger material remains at the higher levels and in the lowest, finest screen you'll get some very small pieces of material that we would feel would be common to an explosive device.'

At the same time, they were interviewing the witnesses and comparing their statements with the facts they had ascertained. They told the interview teams what 'we needed to know about the visual sightings in and around the fire, the colour of the smoke, the colour of the flames, the smells and the sounds that were heard prior to, during and after the fire, all very helpful, each contributing to each other's effort.'

Some of the witnesses confirmed that 'the fire first appeared to begin in this ballroom area where they saw what appeared to be blue flames on the side of the furniture that had recently been delivered and stacked in the back of the ballroom', says Vita. 'Blue flame is fairly uncommon for that type of material, which doesn't normally release a blue flame. We took samples and were able to find in the debris evidence of an alcohol-based accelerant. The fact that people said that the boxes seemed to be burning on their sides is very uncommon. Normally if a fire had started in that area it would start at the lowest part of the box and burn up through the box. That led us to believe that there was a reasonable chance that some accelerant had been applied to the outside of the box. As we conducted our investigation in that area we were looking for material that would be common to that type of a fire and also something that would have the properties that would allow it to adhere to the outside of a box.'

The pile of boxes was big – over six foot high and 31 by 18 feet. The pattern of burning was unusual and thus suspicious. Normally, says Vita, 'they burn from the lowest point outward and upward, but the V pattern of the burn on the boxes indicated that the fire had begun on the outside, on the front edge of these boxes, then burned back up through them.'

If Vita's progress and his words seem ponderous, this is no accident: caution, thoroughness, not taking anything for granted, are especially important in a case as complex as this, in which arson seemed a strong possibility. 'To prove your case,' says Vita, 'you really need the scientific evidence that comes from your laboratory examinations and the metallurgy tests of materials that you recover at the scene to support your assessment of the situation. When we prepare a case for criminal trial it's very important that we not only prove our theory on how the fire began and developed but we have to conclusively eliminate all other possible causes.

'In this case you're looking at accidental causes, such as electrical failures, natural gas problems, anything else that would be common to that part of the hotel. There was no natural gas available to that area but, there were electrical outlets so we had to go through the entire electrical system within the building to ensure that there had been no electrical failure in another area that had been brought into the disaster area through the electrical circuitry or through the walls.

'As we conducted our interviews and our scene investigation,' continues Vita, 'we found samples of aluminium that looked like it had been melted because of the intense heat. It appeared to be like little circles or little aluminium balls and bubbles that had solidified after they cooled. When we sent that debris as well as those pieces of aluminium to the laboratory we were able to determine that the chemical properties of that aluminium were consistent with the chemical properties of aluminium containers of sterno. This is an alcohol–paraffin-based material, often used by hotels to keep food warm. It comes in a small can and, because of the alcohol base, as it's lit there's a flame that comes off the top of the can. At the bottom of the can it's very sticky, pasty, and would apply well and adhere to the side of a box. We found evidence of an alcohol-based accelerant.' They finally concluded that this intensely inflammable material, which was in general use in the Du Pont as in most other hotels 'had been applied to the outside of those boxes'.

At this point the investigators ceased to look at evidence and started to act as traditional detectives. 'As we started to interview the witnesses,' says Vita, 'we found that the people who were closest to the area of origin of the fire at the time they thought the fire began were some of the two dozen or so security guards hired by the hotel after earlier labour problems. One of the guards was a particularly interesting witness in that he got very emotional when he started to talk about the scene itself and what he saw and what he did. As we interviewed him further we found that he said and thought that he did things that may not have been possible to do. He talked about opening doors and going through them from one section of the hotel to another, which wasn't possible because the doors didn't open in that direction. We interviewed him again with hotel staff who were familiar with the mechanics and spatial relationships within the hotel and they questioned some of the things that he said as well. Finally we conducted another interview and walked him through the hotel while we were interviewing him and asked him to show us what he did, what he saw and what he heard at the time of the fire. As we did that he became exceedingly emotional and he was perspiring profusely. Our first thought was that he was possibly involved because he had that type of an emotional reaction to the scene, especially when we walked him into areas where they were still removing bodies and he saw the remains of the people who had been killed. We interviewed that individual continuously for a couple of days. We tried to play to his role as a security guard to get him to help us figure out what happened and he had various theories that he expressed to us on how it had, but the further we got into the investigation he had all the appearances of a classic suspect in the case.

'After several interviews and discussions, we found out that he had not been part of the group that set the fire but that he had seen the people who did and he was fighting within himself to give us the information. Eventually he did, which helped us identify the suspects in the fire.'

When the investigators put their theory as to how the fire had been started with the sterno to the other interviewers, one, says Vita, 'thought back over an interview that he had conducted with a witness who had talked about seeing people in that area. The witness seemed to be squeezing his hand, unintentionally and probably unconsciously, but as he was describing what he saw he was going through the squeezing motion. Now, we instruct our interviewers not only to listen to what's being said but also to examine the body language of the witness. In this particular case the interviewer recognized that there was some inconsistency between what the person was saying and the squeezing motion. Later on we put those two pieces together and realized that there was a good possibility that the sterno had been poured into plastic ketchup-type bottles and from that that the squeezing motion could have been the constricting of that plastic bottle in the expelling of the material on to the sides of the boxes. It was rather interesting how that came out during discussions.'

In the event three of the hotel's employees were arrested and found guilty of arson. But this was no ordinary case and their conviction was by no means the end of the story. To Tom Klem, fire protection engineer, 'the traditional approach in the investigation of a large loss-of-life fire like the Du Pont Plaza is to do an analysis based on the growth, development and spread of the fire and it's really the experience of the investigator and their knowledge of applicable codes, fire barriers and fire-protection features that can be implemented so that was one prong of this investigation.'

But – and here lay the original nature of the investigation into the Du Pont Plaza blaze – scientists and engineers from outside the investigatory profession were asked more fundamental questions. As Klem puts it, 'was a time-frame such as three minutes realistic for the flash-over of the ballroom? We were able to answer that question. Yes, it was, and we were better able to understand the growth, development and spread of the fire through that casino as a result of that. It was the first time in fire-protection history, really, that the practical investigative techniques of cause and origin were married with science, and that joining brought about a more conclusive result not only from the skills that the on-scene cause-and-origin people brought but also the skills and the science that came into the investigation with the scientists.'

The Du Pont Plaza fire provided an ideal opportunity for the scientists at the National Institute of Standards and Technology (NIST) not only to explain why the fire developed so quickly but also to prove the theories they had developed, with the ultimate aim of discovering, in the words of Bud Nelson from NIST: 'Why did the fire get as big as it got and, secondly, why didn't it get any bigger?' It happened, he says at the 'right time. At NIST we'd developed a package of tools that related to the rate of fire growth and production of energy and smoke from burning materials, how this fire reacted when it was inside a particular space, how the smoke built up, how hot it got, what the impact of that heated smoke was on causing other things to burn, and how much energy they released, how much mass or smoke particles they released and then, given this, how they responded and reacted in their environment. Where does the smoke go? How does flame spread across surfaces? How does smoke move? What is smoke? Is it a combustible material itself or is it the final product of combustion?' They had also been working on computing the way fire spread 'and we had tested many in our laboratories here but these are all relatively small-scale, the biggest being the size of a small town-house'. The concentration of investigators at the Du Pont Plaza provided Nelson and his colleagues with enough information to enable them to test whether their theories could be scaled up from domestic size to a major building.

Nelson took with him to Puerto Rico a laptop computer loaded with the equations he had developed. He traced how the fire in the ballroom 'produced a hot gas that accumulated at the ceiling of the south ballroom, getting hotter and hotter until the temperatures in that gas were in the range of 600° centigrade or so, at which point the radiation from that hot body caused other things to ignite, such as the combustible wall between the two ballrooms and the material that they call wallscaping – something heavier than wallpaper and lighter than carpet on the other walls – and also caused enough stress in the glass between the south ballroom and the two-storey foyer for the glass to break out, bringing in a new quantity of air and involving flash-over where everything in the room tries to burn to the extent that it can find air.' All this happened in a mere six or seven minutes.

More important to the scientists was what happened next: 'The amount of fuel being produced by the heating was three or four times more than the amount of fuel that could be consumed by the air that could be drawn into that room. So the fuel got carried in the smoke in two directions, into the north ballroom, because by now the wall between the two ballrooms was failing, and out through the foyer. This now fuel-laden flame and gas filled the upper portion of the foyer area exposing both the main lobby and the glass windows that separated the foyer from the casino. This fuel then moved with the smoke into the foyer where it found more air and burned more in the foyer but didn't burn it all: it was still a black sooty flame. This broke the glass into the casino and carried the unburned fuel into the casino. As it found air it travelled the length of the casino in twenty to thirty seconds as a big flame front, catching a number of people, who were instant victims of the fire, and trapping others. In the casino the people were bathed with flame. The post-mortems on them showed very low carbon dioxide – their lungs were burned as might occur if they'd been hit with a flame-thrower. They were just engulfed in flame as this flame front came across. It then broke out of the far end of the casino through the window over the pool and bar areas, trapping two individuals at a table down by the poolside bar, finally terminating as a fireball out at the far end over the pool. At that point the fire found enough oxygen to complete its combustion and then burned out all the areas behind.'

That the model was very close to the physical reality was clear from the story of the glass partition, between the stairway to the casino and the main lobby, which pointed to something that no one had noticed. 'When we ran the model,' says Nelson, 'because there was no evidence in the fire investigation we had assumed that there was no partition or any separation between the foyer and the main lobby of the building. When we got back and started exercising the model carefully we came to the conclusion that had there been no partition there the fire in the hotel lobby would have been much more severe in the early stages and possibly many more people would have been trapped. We wanted to find out what had stopped the fire from going in there. Exercising the model told us that most probably there had been a partition there that we were not aware of, a partition with a door in it that allowed the smoke to go into the hotel lobby but restricted it so that the flame did not really follow it.

'Early in February 1987, I went back down to San Juan, dug in the rubble and found the track where a glass wall had been. Then I started asking questions about it and found some people who said, "Oh, yes, we had a glass partition there with a doorway in it." Later we found some video of a wedding which showed this glass partition and to me it was very interesting that the model had demonstrated something that we'd all missed by looking at it physically.'

To Andy Vita, 'The fire had special significance to the fire-investigation community. It brought together not only the investigators from the fire-suppression areas, fire investigation, but it also brought representatives and experts from the academic and scientific communities. They talked about and reviewed some theories that they had been developing on fire modelling and progression. When we go and testify in court it's very important that we can convince a jury of the facts we have found and that we have scientific evidence to support our statements. This fire provided the forum for a number of experts to come together and really analyse theories they had previously developed to the point where fire-modelling evolved from that investigation. Modelling that we use today really allows us to evaluate the science of how a fire developed, how it moved through the structure, and in this case how it was ultimately responsible for the death of ninety-seven people.'

They developed 'computer modelling that really helps us analyse how fire develops from heat and movement, how the gases are released and developed, how temperatures reach certain levels. Often times we have to try and estimate certain temperatures to determine the effects on the particular materials within a building, what products of combustion are released when those materials burn – are they toxic gases? It helps even in developing furniture, wall-coverings and flooring so that we can create materials for the interiors of commercial buildings that will not release those toxic chemicals.

'They actually put together a minute-by-minute sequence of the fire and how it developed so that we could match people's witness observations to the actual science of the fire, that if the fire began in this area at this time within so many minutes it could conceivably move from that point to another point.'

In Bud Nelson's words, the Du Pont Plaza investigation provided 'a strong basis for demonstration that the scientific modelling we developed here at NIST was useful, could be applied in the field and could be used by the practising engineer or fire-investigation professional. The work at Du Pont Plaza opened the door between traditional fire investigation and fire science. That door has stayed open and communication has been growing ever since, particularly between ATF and fire science, and between the entire fire-investigation community in the United States and, to some degree, elsewhere. Since then I have gone to various places in the world giving lectures on the use of models in fire investigations – Italy, England, Japan – and it has become almost a standard practice. Today, the fire-investigation community recognizes, appreciates and hopefully uses fire science in investigations.'

# 3
# Two Fires, Two Traumas, Two Triumphs

There are probably no new lessons to be learned, it's just a reinforce-
ment of all lessons that we have seen before that, unfortunately,
replayed themselves at this incident.

        Ed Comeau, on the fire at a disco in Gothenburg, 1998

In the past few years, two fires, in central London and in Dublin, shocked
their local communities to the core.

On the evening of 18 November 1987 a small fire was spotted on the
escalator leading to the Piccadilly line platforms at King's Cross station, one
of the busiest and most complex on the whole London Underground network. The fire and smoke spread to the escalator shaft, the ticket hall and
the passageways leading from the ticket hall to the street and to the main-
line station. Thirty-one people died and many others were severely injured.
It was probably the worst such accident to have hit the underground system
this century.

It was also totally chaotic. Even after the fire broke out, 'People were still
getting off trains,' says Ian Rogers, King's Cross Station manager, 'although
attempts were being made to prevent that happening, and people were
coming into the ticket hall area. We had staff at the foot of the escalators at
the Piccadilly line who were diverting people up the Victoria line escalators
and there were people in the ticket hall when the fire developed to a point
where it flashed over and the ticket hall was engulfed in flames and very
thick smoke.'

When Dave Halliday, the first fire investigator to arrive examined the
devastation he was surprised by the degree of damage at a high level. The
ticket barriers at the top of the escalators had been destroyed, and the
booths were badly damaged but 'looking across towards the Piccadilly line
escalator, what I found was that the worst damage in the concourse was

concentrated around the opening to the escalator. I then decided to concentrate on the escalator and looking down I could see that the fire damage extended approximately two-thirds of the way down the escalator itself. There was a very sharp demarcation to the damage, part-way down, particularly along the ceiling where it appeared almost like a line of burning had been cut into the paintwork. I decided, using the guiding principle common to most fire investigations in which the lowest point of burning is usually where the fire started, to concentrate on that area.

'At the bottom of the escalator, it was quite clear that there wasn't any burning at all. Any person standing at the foot of the escalator during the fire would have been perfectly OK but looking up from there you could see clearly demarcated damage so I decided that I was going to concentrate on that low area of damage on the left-hand escalator and probably looking more towards the right than the left because that was where the greater degree of burning had taken place, to the stairs and the skirting board.

'Looking closely at the escalator steps, particularly where they ran alongside the skirting board, I found that there were quite big gaps between the edge of the steps themselves and the skirting boards in a number of places. They were so large that I could insert my fingers into the gap and get my hand virtually half-way down. These gaps proved very significant because it was through them that we believe that the fire was actually introduced on to the trackway of the escalator by a burning match.

'To allow free movement of the escalator steps there must be a small gap between the steps themselves and the sides of the escalator. This gap is often filled with a metal channel. However, on the Piccadilly line escalator some of the metal channels were missing and probably over the course of years a certain amount of movement had taken place. As a result there were very thick gaps sometimes between the sides of the steps and the skirting boards.' In addition, 'a thick layer of grease had built up between the wheels due to the running action of the escalator. Anything dropping through the gap would land straight on top of the grease so if a lit match went through, it would ignite the accumulated grease. This might then burn the skirting board, leading to the small patterns of damage that we found when we dismantled the escalator. What we then needed was some explanation of why there was this pattern of small burns on the right-hand side of the up escalator, which didn't seem to be reproduced on any other part of the escalator system. I believe the reason for this is in the pattern of usage of the escalator. If you're standing on an escalator going up out of the station you might want to light a cigarette. If you're stationary it's easy to strike a match, light your cigarette and then, in your own eyes, safely dispose of the match by throwing it forwards and downwards, away from the passenger standing above you on the escalator. It's not so easy to light a match if you're walking up the escalator. On the down escalator, first, you're entering an underground system where smoking is forbidden and, second, if you throw a match you'd be throwing it down on to the passenger in front of and below you so I think you would be inhibited from doing this. It did seem that all these small burns might be the result of passengers lighting cigarettes and disposing of matches as they came up out of the underground system.' (Today, smoking is forbidden throughout the underground system.)

'A great deal of timber had been used in the construction of the public side of the escalator: the stair treads were of wood, the stair risers were of wood on a steel frame and the sides of the escalator were steel panels covered with a wood veneer. None of this material would be readily ignitable, though – it would need some kindling to start it going and I felt that there was nothing in the public area that would be sufficient to start the fire burning so I turned my attention to the underside of the escalator. There were other reasons for doing this as well, one being that we'd had early eyewitness evidence to suggest that there had been a fire burning under the escalator before the first flames were seen on top.'

Once he'd realized that the fire had not started above the escalator steps Halliday went beneath the escalator. 'I found myself disappearing down a rather deep, gloomy tunnel. There is an extremely narrow walkway between two of the Piccadilly line escalators, it's very dark and hot down there. I was having to use a hand torch all the time I was working there and one of my colleagues had an encounter with one of the local inhabitants – a rat.

'When I looked at the side of the escalator from the underside I found that virtually all of the combustible materials had been consumed, from the level of the trackway upwards. However, the return side of the stairs, which could be examined by walking down underneath the escalator itself, showed no signs of burning. This indicated that the escalator had stopped before the fire had become established in the public areas, which meant that those stairs which had travelled past the top of the escalator and were on the return when the escalator stopped were never exposed to fire. On the lower section of the escalator skirting board on the right-hand side there were fifteen small areas of charring. A lot of these were hidden behind metal struts and stanchions.

'Early witness reports had indicated that the fire on the escalator was stationary, passengers had seen a fire below the level of the steps. This meant that really the fire could only have been burning on one area of the escalator, the non-mobile part that had fuel on it.

'Most of the escalator structure was non-combustible; it was made of concrete and steel. The only burnable materials were a thick layer of grease, which had built up between the wheels on the trackway, and the fluff and other objects impregnated in it. There were matches and bigger items, such as pens, pencils and, lower down the escalator, I even found a complete newspaper there.

'Now I had the information I needed. What had happened was a passenger lighting a cigarette had dropped a match, which had slipped down between the steps and the skirting board and landed on the grease and fluff below. Unfortunately the match was still alight when it landed there, the grease and fluff ignited and a small fire developed on the trackway. Passengers reported this fire, they were being carried past it by the escalator, but when it became more serious the escalator was stopped and from that point onwards the fire spread into the public areas and eventually got large enough to destroy the entire upper station concourse.'

Normally investigators work on the principle that a fire usually burns upwards not downwards and that the point at which it started will show the most damage – because this is the point at which the fire has been burning longest. At King's Cross the opposite was true. Above the fire damage even the advertising posters on the walls of the escalator shaft were untouched. But the biggest problem facing the investigators was the way in which the fire had spread up the escalator and burst like an exploding bomb into the ticket hall. They could not explain it.

In Britain the Health and Safety Executive (HSE) has the responsibility for safety on railways. Their laboratory is housed on the site of a First World War munitions dump high up above the little spa town of Buxton in the Derbyshire Hills, and they came under intense pressure to explain this national tragedy. They had two objectives, says Keith Moodie head of fire/explosives. 'One was to establish how and where the fire had started, and the other was to explain how it had travelled so rapidly up the escalator and into the booking hall.'

Their tests agreed with Halliday's theory as to how the fire had started. They were then faced with what Moodie describes as 'the difficulty that eye-witness statements clearly stated that the fire had spread rapidly, and people had been caught off guard in the booking hall by a sudden event. The question was, how could the fire, which had seemed fairly innocuous, have developed so rapidly and spread into the booking hall?

'We rebuilt five or six steps of the escalator to full size, using wood taken from unburnt sections of the other escalators. We dropped a match down the side and on to a simulated grease layer, which caught fire fairly rapidly. The fire spread under the skirting board and underneath the escalator. When the fire had spread on to the visible side of the escalator it spread slowly across the treads and risers of the escalator and also up the balustrades on the sides. Eventually the flames from one side reached across to the other and ignited the other side. At that point the whole escalator became involved and we were able to use this to assess how the fire had spread across the escalator but also to give us information on the heat output from the fire and the power it was developing.

'We now had to establish how it had spread so rapidly *up* that escalator, about twenty metres in less than a minute. To do this we had already con-tracted the people at the Atomic Energy Authority at Harwell to undertake a computer simulation of how a heat source placed in the trench of the escalator would behave?

The HSE knew, says Ian Jones, CFX chief technologist, Atomic Energy Authority Technology, that Harwell 'had a great deal of experience in pre-dicting the flows of fluid, of heat transfer and some of the interesting mech-anisms for aerodynamics. We were also involved in developing and pioneering the new technology of computational fluid dynamics and they felt that this technique might be able to help them understand the complex aerodynamics of the flow in the escalator tunnel and in the booking hall. Previously people had not been able to cope with the complex geometry of such areas and had looked at the fire spread in fairly rectangular rooms. The situation in the escalator was different so they wanted us to use these new techniques to give them some understanding of what might have caused the very rapid spread of the fire.'

Once they'd been properly briefed, the scientists, says Jones, had 'to set up a grid system for the region of interest. We set up little bricks, like Lego bricks, connected all together. In each of these bricks we would be solving the fundamental equations of fluid flow, they would all be connected together. Then we'd have to solve the equations in these tends of thousands of cells for many time steps to watch the development of the fire. At that stage it required very large computing power and we were lucky the world's largest super-computer was available to us, the Kray 2. We used it to explore new avenues, new modelling techniques that could give us a greater understanding of the fire dynamics.'

First they set up a simple model 'which consisted of the complex geometry of the escalator, the escalator trench, the balustrades, the booking hall, the ticket office inside the booking hall and the passageways around, to the orbital passageway and down to the Victoria line'. In a mere eight days, working twelve hours a day from the engineering drawings, Suzanne Simcox modelled the underground station.

'First,' says Simcox, we 'defined the geometry of the tunnel and the booking hall to the computer. Then we wrote some software to join the two pieces together. We then split that geometry up into lots of little cells, like Lego bricks, and the computer joined them together. In addition we speci-fied the fluid – air – and the boundary conditions, or what's happening at the foot of the escalator shaft and at the edge of the booking hall. The creation of the grid took us further than we had gone before with the tech-nology at the time. We were used to dealing with one thing at a time, like the escalator tunnel or the booking hall. Joining those together was quite new for us and the calculation was pushing to the limits of what the software could do at the time.'

The work, says Ian Jones, 'threw up the most difficult computations we'd ever done. It was a new technology that we were pioneering for other reasons, for the nuclear industry and for the commercial work of the laboratory, but this really pushed it forward because it was being driven by real needs, real time scales and we had to do something because we wanted to help.

'At first,' he continues, 'we were surprised at the results because they seemed not quite right. They were showing that the hot temperatures were near the floor of the escalator and everybody knows that hot air rises. In a classical fire you see the smoke plume rising and the fire spreading across the ceiling. We'd expected this kind of thing to happen but the simulation showed that the fire, the heat, was sticking down low in the escalator trench, and we couldn't understand that. We thought at first that we'd got a bug in the software, but when we became convinced that what we were doing was right we talked to some experts. Gradually we understood the physics of why the flames could lie low in the trench of the escalator. It's the so-called colander effect, whereby jets stick to walls. If people had seen a jet sticking to a wall that was vertically upright they would not have been surprised, but because it was at an inclination of 30 degrees and the flames were sticking they were surprised. Once we understood this mechanism we were reasonably confident of telling other people about it.'

Then the HSE commissioned a small-scale experiment at Edinburgh University with Dougal Drysdale and his colleagues. 'They took a four-inch diameter tunnel made of cardboard and set fire to it', says Jones. 'The flames stuck to the floor of this little tunnel and from then on I think we knew we really had got something. We were able to say that the flames and the hot gases ahead of them could preheat the wood in the escalator trench and then instead of just burning cold wood you were burning very very hot wood. This explained why you could get a very rapid spread of the fire from a wood fire when people expected wood to burn relatively slowly. That was the key to the rapid spread of the fire and the development of the flash-over.'

Keith Moodie takes up the story: 'The results indicated a possible mechanism whereby the fire could have spread more rapidly and may explain the sudden burst of flame into the booking hall observed by the witnesses. They showed how the heat from a source placed in the trench of the escalator stayed low instead of rising. This was due to aerodynamic effects but it gave us good guidance as to how the fire might have behaved. The hot gases were preheating the wood and the fire spread rapidly through that preheated wood at an ever-increasing rate. As the size of the fuel bed – that is, the wood that was burning – increased so did the flame extension and hence the rate at which the fire was spreading up the escalator. You had a very rapidly moving flame front coming up the escalator. It probably took less than thirty seconds to travel the length of the escalator involved in the fire. It would have meant that at the top you'd get a sudden jet of flame entering the booking hall and this would travel across the booking hall roof. It would also have been accompanied by smoke and a very rapid increase in temperature.

'Our first reaction was surprise and questioning of their accuracy but through discussions with Harwell we all satisfied ourselves that the input data was correct. It was unique in that it showed that the direction of the flames was along the trench of the escalator rather than up into the air and it was certainly that aspect that took a lot of time to clarify and understand. We then commissioned colleagues at Cambridge University to advise on how we could model this.'

They went on to build a third scale replica of half the Piccadilly line escalator system plus part of the booking hall into which it ran. 'It was an enormous undertaking for our staff, particularly our workshops people and our designers,' says Moodie, 'in that they had to work against the clock really to get this model built and in place and tests under way all within a month. We instrumented the model so that we could measure the temperatures and the rate at which the fire was spreading up the escalators and we had a computerized system back here to log all the data from the instruments. We videoed and viewed the tests from down below and from cameras positioned in the booking hall and at the side of the escalators.'

Suzanne Simcox was at the first experiment. She was 'standing at the bottom of the escalator trench. When they set light to it the flames stood up to start with, but as they moved across the trench and took hold they lay down in the escalator trench and someone at the top shouted, "It's just like Harwell said it would happen," which was nice to hear, standing at the bottom!'

'What the experiment showed,' says Moodie, 'was that a fire would have taken eight or nine minutes to develop in the trench of the escalator, but then it travelled up the escalator in less than thirty seconds. It would have meant that anybody at the top would have been overtaken by a jet of flame. On our models, and on the experiments and calculations, we showed that temperatures were notably rising at twenty or thirty degrees per second in the booking hall. It went very rapidly from a survivable atmosphere to one that was unsurvivable.'

The work, says Ian Jones, 'really gave people a new insight into the reasons why the fire had spread so rapidly. Before that people were concentrating on other reasons for a rapid spread. A lot of people were looking at the effect of the paint. They might have been able to explain through paint burning why there was a rapid spread of the fire but the theories didn't really fit together. Once they'd seen that the wood could burn so quickly and so fast because it was preheated, that gave them all the clues that they needed to look down at the floor of the escalator rather than looking up at the paint. It was that clue that the computer modelling gave them, to look down and things will be explained, rather than the fairly obvious way of looking up to see the ceiling and what could have happened to the ceiling. We gave them that clue and from then on people understood and there was not really any discussion, any argument at all about why the flash-over occurred. The effect became known as the trench effect, it was a new phenomenon in fire dynamics.'

The fire led not only to a new understanding of the dynamics of fire, but to far-reaching changes in the organization of emergencies throughout the underground network. 'We now have a co-ordinated emergency plan of which the emergency services are aware,' says Ian Rogers, King's Cross Station manager, 'so the fire brigade and the police knew what action we would be taking in another such situation. All our staff are trained in how to evacuate customers and what points they should go to within the station. In a typical evacuation I would have staff manning all of the gates leading into the station, which would be closed to entry but allowing people to exit. All train services would be non-stop through the stations as soon as people had been evacuated from platform level. Anybody in the area would be cleared out immediately to street level, and we have a rendezvous point where we would meet the emergency services and bring them into the station. We have a plan to which the emergency services have access which details the whole layout of the station. We can explain to them exactly where the emergency is and those plans are available at street level. In addition to that all our staff are in touch with our control room by two-way radio and we have an integrated public-address system covering the whole station so that by the touch of one button we can broadcast an evacuation message that will reach every area of the station.'

If King's Cross was traumatic for Londoners, then Dublin was devastated when 48 people died and 128 were seriously injured when 800 people were gathered for a Valentine's Day dance at the Stardust disco in Dublin in 1981. It was the worst single disaster the city had ever seen.

'The Stardust was really the only big gathering place that there was for people in Artane [a Dublin suburb] in those days,' says survivor Father Heber McMahon. 'Various football clubs would have smaller pavilions and recreation areas but it was *the* place. You mentioned the Stardust and people would associate it with not just the discos but with concerts. I knew the building because I was part of a group of priests who did fund-raising, and charity shows, performing in concerts, and we had been at the Stardust frequently.'

The fire had started in what was called the west alcove, a separate area from the main disco floor with seats in tiers and walls on three sides covered with carpet tiles. When the fire was first spotted no one was particularly concerned – survivors described it as small and controllable – but within just two minutes everything in the Stardust was ablaze, seats, walls, ceiling, floor, tables – even the metal ashtrays. The evacuation was chaotic. 'There were chains on some of the exits,' says Father McMahon. 'It was possible to open the crash bars from outside and get in so they had chained them.'

The fire left a major mystery: how did it spread so rapidly?

When Seamus Quinn, a Dublin fire investigator, reached the scene at 2.30 a.m. he found a sight familiar to all investigators. 'There was an outer cordon surrounding the site,' he says, 'and outside that cordon there were small groups of people in total panic, in shock, hysteria, some trying to comfort those who were crying uncontrollably. Inside the cordon the emergency crews, fire brigade, ambulance service, they were all going about their normal work. Inside, the fire brigade were damping down still burning areas of the site, others were searching for bodies and the electricity board were cutting off the power supply to the site.'

Despite the chaos and hysteria his professional training ensured that he immediately started on his normal routine. His first reaction was 'that this was total devastation. The club itself was totally burned out, the seating, the tables, the carpet wall tiles. The suspended ceiling and the lighting had been pulled down. The area was like a sauna when I went into it – the heat was devastating and the fire brigade were still damping down so steam was rising from the debris.

'My biggest problem as a fire investigator is to try to establish the fire seat, where the fire started. I have to find the cause of fire, be it malicious or accidental, and to this end I must look to outside people, the fire brigade. What did they see when they arrived at the fire? What did the witnesses see? What did people driving by see? Who were the first people to see this fire? I know I'm going to be very dependent on first witnesses at the scene, what the patrons within the scene saw, how those statements are taken and the detail into which statements went. I'm going to have to eliminate [individual cases]: whether electricity or careless smoking was the cause of fire or whether accelerants were used.

'Did the fire spread through the roof at an early stage? That will tell me a lot. Did the fire brigade note the colour of the flames when they arrived?

That will tell me a lot also, as will the amount of smoke that's being produced. It will tell me what materials were in that building. But I depend an awful lot on the reaction and the recollection of first witnesses at that scene, and on my own experience as a fire investigator. I look to the depth of char of timbers, which will tell me, as a general rule, how long the fire was in progress, and the colours of metals will tell me a lot too – but my initial reaction at that scene was of awe. And where was I going to start?

'The witnesses that night described the fire as a small flame dancing along the top of the seats, perhaps some three or four inches high. Others described it as a coat on fire on the back of one of the seats. Another described it as a quite substantial fire half-way up the rear wall and involving the carpet wall tiles. Others described it as almost reaching the ceiling but they all put it at the rear wall of the north banked area.'

But they found 'nothing to say that there had been a coat there. We found no material, we found no buttons, so we had to eliminate a coat being on the back of the seat. We also heard through statements that there was a sack of rubbish up near that area. In that rubbish we would have expected to find bottle tops or beer mats, but nothing like that was found in any of the debris.

'At that stage, I could not comprehend why this fire was not put out. I carried out some small tests in the Garda technical bureau in which I set fire to a seat and, with the seat totally ablaze, put it out with approximately two gallons of water. One of the bouncers actually saw the fire and went for an extinguisher, yet when he got back it was uncontrollable. I could not understand this.

'The fire had started in a banked-tier section with seating right across it. When the seating took fire it would have spread via the carpet tiles to ceiling level. The ceiling was constructed of asbestos tiling so would have reflected down the heat produced and the hot gases on to the tops of the seating in the banked area. So when the first row of seating and carpet tiles went on fire the heat would have been reflected down upon the rest of the seating. This would have raised their temperature to their own ignition temperature and they would have ignited simultaneously without the aid of a flame or a spark.

'When we tested the debris in the forensic science laboratory no accelerants were found so that eliminated one source of fire. When we read through the witness statements, we saw that witnesses had seen the fire on the back row of seats. We had eliminated the coat and the bag of rubbish as potential causes, and after that we were left with the possibility that the fire had started not on the back row of seats but on a row of seats in front of it. We took three rows of seating and set fire to the centre of the three rows. The fire spread neither backwards nor forwards but the centre row of seats burned out, so that eliminated the suggestion that it could have started on a row forward of the back row of seating.

'Experts eliminated an electrical cause, so we tested to see if a lighted cigarette could set fire to the seating. The answer was no, and we were left with only one possibility after that, that the fire was started by a person setting light to a slashed seat with a match or a cigarette lighter.'

With their limited resources the Dublin investigators could go no further. They called in the UK Fire Research Station (FRS) which has access to one of the most extraordinary facilities in the world of fire investigation: an enormous hangar at Cardington, near Bedford, built in the late 1920s to house the ill-fated airship, the R101 that, symbolically, caught fire on its maiden flight. Its size enables the investigators to build full-scale replicas – as they did in the case of the Stardust disco fire.

Not surprisingly the FRS proceeded in the same systematic way as Quinn had. 'There was a suspended ceiling and floor covering on the steps,' says Bill Malhotra, who was in charge of the investigation, 'so the first thing was to try to identify which materials were involved. Most of the carpet tiles inside the disco had been destroyed but out in the entrance hall some were still in place so that gave an idea about the type of tiles used. Seats within the disco were destroyed but next door to the disco there was a restaurant and a pub, which had identical seats. As far as the ceiling was concerned, we were told whose system it was and that it consisted of suspended ceiling tiles. The floor covering was of a plastic fibre type. These materials were identified and we arranged to get samples down to the FRS laboratory to look at them individually.'

None of the materials seemed unduly inflammable but they confirmed suspicions first voiced by Seamus Quinn: 'when we slashed the PVC chair covering and exposed the PVC foam', says Malhotra, 'it was possible to ignite that with a small flame. Once the foam caught, the fire spread progressively along the seat to the back and this involved the whole seat so that the whole seat could burn, producing long flames travelling upwards.'

They increased the size and scope of the reconstruction experiments, discovering that the carpet tiles helped to nourish the fire but that the crucial point in terms of the severity of the fire was that the ceiling was so low. 'Once these seats were ignited,' says Malhotra, 'there were bigger flames under the ceiling which heated the seats in front to a much higher level so that the level of radiation on those seats was high enough for spontaneous ignition to take place. That occurred and the fire spread to the next tier and the tier in front of that. As far as we were concerned that established that we had found out the mechanism responsible for fire spread in the west alcove, not only along a tier but from tier to tier.'

They finally concluded that 'the fire was started in one of the seats next to the wall to the aisle [a finding they confirmed] by burning three or four sheets of crumpled newspaper underneath the seat. The seat became involved, the fire from the seat spread upwards to the ceiling tiles and below the ceiling spread sideways to the adjacent seats towards the side wall and three or four seats became involved at the same time. In one or two minutes, when the fire had advanced sufficiently for the fire in the tiles near the ceiling to start spreading to other tiles across the aisle, those tiles started to drip flaming drops and caused the seats across the aisle to start burning. So we had the whole rear tier burning within one and a half to two minutes of the start of the fire.

'Very quickly after that the heat was sufficient to cause the spontaneous ignition of the seats in the tier in front so that within the next ninety seconds that tier became involved. It was a matter of a few seconds only after that before the whole of the five rows of seats started to burn. The flames started to come from outside the rig accompanied by lots of smoke, so that the whole thing was on fire. It was as if we were repeating the fire in the west alcove.

'The seats would burn but they were not highly inflammable because of the PVC covering. They would not ignite from a lighted cigarette or from a lighted match unless, of course, the covering was damaged and the foam was ignited. By itself a seat would have produced heat but the fire would not have spread to other seats in the vicinity. When the carpet tile was present behind the seat, the fire from the seat was able to ignite the carpet tile and in combination the flaming was increased in intensity and spread upwards along the wall. The tiles had another feature, that they produced molten drops which fell on the adjacent seats. If the seats had been heated they ignited.'

But 'had there been a tall wall with carpet tiles and seats and no ceiling or a very high ceiling, then the flames would have gone up on the wall. Seats in one tier would have become ignited but there would not have been sufficient heat to heat the seats in the front tier and for the fire to spread to the front. It needed a third component, and this was provided by the low ceiling, which allowed the flames to collect under it, radiate heat along to the seats sufficiently for those seats to become ignited spontaneously. It was this combination that was necessary for this fire to spread so rapidly.'

When Quinn saw the experiments, 'I was amazed,' he says, 'at the rapidity with which the fire spread from an initial small flame started by a sheet of newspaper beneath the seating on the back row of the rig to the conflagration that occurred some few minutes later, and at the rapidity of the production of volumes of toxic black smoke. I was amazed at the speed with which the seating seemed to smoke and then catch fire almost simultaneously. The test proved that the fire did start on the rear row of seats in the Stardust disco. It also proved that the carpet wall tiles assisted the spread of fire. It proved that the emission of smoke was huge in volume and may have obstructed the view of the patrons in the disco. It proved the fast pace of the spread of fire from the back row of seating to the front.' Intense heat radiated downwards and everything else in the club burst into flame. Within five minutes everything that could burn had been consumed.

Discos – often overcrowded and gimcrack constructs of inflammable materials – are notoriously liable to disasters like the one at the Stardust, which are the more tragic because the victims are so young. This was the case in a fire that destroyed a hall in Gothenburg in western Sweden in 1998. The only difference between that and the fire at the Stardust is that the Swedish authorities have been less forthcoming with their explanations than the Irish. 'It was a fire that happened on the second storey,' says Ed Comeau, an American investigator who examined the blaze. 'It was being used as a disco by teenagers. There were too many people in there and since the fire occurred in one of the stairwells, there was only one means of egress from the building.' To make matters worse, 'One of the stairwells where the fire originated had a large amount of furniture piled into it. This created two problems: first, it was a blocked exit, but there was also a fuel load in the stairwell.

'A lot of the people tried to move through there very rapidly, it became backed up, people started jumping out of the windows from the second storey and hurting themselves – they were six metres from the ground, so they were breaking legs and getting injured. Another large group of people tried to move into one of the side rooms inside the hall thinking they could take refuge from the fire in there, but indeed they couldn't because about another twenty people were found dead inside that room. The bottom line was that 64 people died in this fire and about 150 were injured.'

And the lessons? 'One is over-occupancy – too many people in a place they just couldn't safely escape from. Another is a lack of sprinkler systems. If there had been a sprinkler system in there [it was not a legal requirement at the time] this fire would not have had the same tragic outcome. A fire-alarm system perhaps would have given them earlier notification of the incident so they could have reacted properly. Also there was the combustible fuel load – and the human behaviour aspects. We've seen a number of these things in previous incidents that recurred were with a tragic outcome.'

# II

# BUILDINGS AND CONTENTS

# 4
# Tombstone Technology

When we think back to the words of George Santayana that those who forget history are condemned to repeat it we realize that if we don't learn lessons from the Coconut Grove then all those people who suffered and all those people who died did so in vain and we don't want to do that and fortunately we have learned some great lessons from this particular event.

Casey Grant, assistant vice-president, NFPA

One of the saddest and most frequent lessons learned from fires is that some aspect of a building itself was to blame. When an historic building, a church or stately home is damaged or destroyed by fire we mourn its fate. But it's also a good, if macabre, opportunity to realize just how subject to fire these buildings can be and just how far we have come in preventing fire by using safer materials and construction methods. The special problems of historic buildings, as noted in *Heritage Under Fire*,* include 'exposed timber floor structures, walls lined internally with combustible materials such as wood panelling, or externally with weather-boarding and roofs of shingle or thatch. There may be interconnecting voids behind panelling and wall linings or undivided roof spaces through which fire and smoke can spread quickly and undetected.'

As with any other type of fire, the causes are many and various. In July 1984 York Minster was severely damaged when lightning struck either the lead roof or a strip along the ridge of the roof, igniting the dry timber beneath (the blaze remained unnoticed for a time because the fire detectors were positioned well below the apex of the roof, leaving a

---

*Heritage Under Fire: A Guide to the Protection of Historic Buildings.* Fire Protection Association London, 1993.

space in which smoke could collect undetected.)

Damage can often be traced to the – usually inadvertent – hand of man or woman. When Hampton Court Palace was devastated in 1986 the cause of the fire was traced to a cigarette end carelessly dropped by an old lady who had occupied a grace-and-favour apartment in the Palace. But such buildings are often most at risk when modern building tools are used in an inappropriate setting. *Heritage Under Fire* mentions a fire at a Wren church in 1988 in the City of London started by a blow-lamp. In 1989 the seventeenth-century Uppark House was badly damaged after roof timbers had been set alight by the heat from an oxy-acetylene torch. Five years earlier, a blow-lamp being used to burn paint off a window-frame set light to Heveringham Hall in Suffolk. A blow-lamp was also responsible for the fire that damaged a Victorian grammar school in Aberdeen in 1986.

In the past the level of fire protection ranged from little to nothing. A British Fire Prevention Committee was formed as far back as 1897 but, as Gavin Weightman remarks in his book on the great fire that devastated London's Smithfield meat market in 1958* 'Brigades did not have the legal powers they have today to enforce regulations.' Indeed, he emphasizes, 'The single most significant change in the work of the Fire Brigade since the last war undoubtedly has been in the dull but effective business of fire prevention.'

Over the past century or more governments throughout the world, often spurred on by disasters, have built up a reactive approach to human safety through the development of prescriptive building codes. Thanks largely to political pressure from opponents such as builders and property owners, the lessons learned from tragedy have taken years, often decades, to be translated into regulations.

One of the best examples of the horrors required to effect change, and the time it takes to carry it out, is the blaze that gutted the Triangle Shirtwaist Factory in the Asch Building in New York City in 1911. In it, 146 female garment workers died when fire spread from the eighth to the tenth floor, the worst industrial disaster in American history. At the time New York's building codes provided that only buildings of more than eleven storeys should have fire-resistant stone floors and metal windows. The Asch Building was lower so had wooden floors, wooden windows and wooden

*Rescue*, Boxtree, London, 1996.

fittings. But that was only part of the problem. Investigation of the fire soon revealed that the Asch Building was a death trap: exit doors from all floors had been built to open inwards so the pressure of people trying to get out prevented them being opened. What's more, some of the doors were locked. The building had only two narrow interior exit stairways. As these filled with heat and smoke, many turned to an exterior fire escape. It was made of metal. Heated by the fire, it pulled away from the building, sending people flying into the courtyard below. Many of those trapped by the fire were left with an impossible choice: either to die in the fire or to jump to their deaths. One of the many lessons drawn from the fire, as Casey Grant of the NFPA (see below) points out, was that 'an outside fire escape is not necessarily an acceptable means of egress as the fire in this case burned through the windows and spread up the exterior of the building, blocking the escape on that fire escape as well as the inside stairwells.'

The fire and the death toll caused a public outcry and started the push for the first fire-safety codes in the US. In the words of Casey Grant, until the Triangle fire the major emphasis was 'on the protection of property and cities, to prevent whole cities from burning down in these sweeping conflagrations which were occurring at the time, and although there were other large life-loss fires prior to the Triangle Shirtwaist fire it wasn't until that particular fire that we saw a dramatic shift in people's attitude, that it isn't just property we need to protect, but people. Safety of people's lives is paramount.

'Directly out of the Triangle Shirtwaist fire came the creation of the National Fire Protection Association (NFPA) committee on safety to life, and the building exits code, which today is the NFPA life-safety code.' Even so, it took sixteen years before the code was drawn up in 1927 by the NFPA. It dealt with the need for commercial buildings to have adequate and properly designed exits and emergency fire escapes.

Codes like this tend to turn into compilations, with new requirements piled upon old ones, making them increasingly inflexible and creating resentment and willingness to cheat among builders, architects and their clients. Worse, there are so many codes, covering housing, electrical circuits and products, plumbing, air-conditioning, ventilation and lifts, that architects and builders find them difficult to comply with, and authorities find them even more difficult to police.

Yet the basic ideas employed to avoid fire and to protect a building's occupants from fire are simple: compartmentation limits fire spread to a given room or a given area; sprinkler systems control or suppress a fire; an alarm will notify the people in the building of the fire so that they can react and escape.

Even today political pressure has ensured that picturesque but dangerous untreated wood-shingle roofs are still legal in much of the western United States. And even when building codes are adequate and properly enforced, they exert no control over the products used to decorate or furnish a building. It is even more difficult to ensure that rebuilding work does not turn an inherently safe building into a fire trap, which has been a major element in many of the fires described in this book – like those at Leo's Supermarket (Chapter 8) in Bristol, the Du Pont Plaza in Puerto Rico (Chapter 2), the Beverly Hills Supper Club (Chapter 13) and the MGM Grand in Las Vegas (Chapter 11).

Moreover, many protection systems, like the multiplicity of fire doors in modern buildings, are regarded too often as unnecessary and inconvenient. Yet they work. Kathleen Collins, a wheelchair-bound lawyer who survived the World Trade Center fire, testifies to this. 'There really wasn't smoke in the actual office areas, it was out in the hallway, so we moved back into the office areas. I guess I used to complain about the fact that we had so many doors, but they worked to our advantage that day because they dramatically cut out any type of smoke coming into the office area and we waited there for quite a while.'

One of the most shocking revelations to me when I was researching this book was that so many of the fires described by the investigators had resulted from lack of a proper sprinkler system. Very often the rules covering their installation are bizarre – in the USA they are mandatory only in hotels used by employees of the federal government travelling on business. However, the simplest sprinkler system would have put out the original fire in the ballroom of the MGM Grand within a minute or two; it would have saved the Garley Building in Hong Kong; and the massive fire at the Meridian Plaza office building in Philadelphia (both Chapter 5) was finally put out by ten tiny sprinklers that had been fitted not by the building's owners but by the tenants. When Ed Comeau, chief fire investigator, NFPA, examined the disco fire at Gothenburg he found it 'frustrating when you see the same problems just resurfacing time and again. If there had been a sprinkler system in there, it would have been a non-event. Essentially a sprinkler system probably would have controlled the fire if not put it out and given everybody plenty of time to get out of the building. If there's one thing that could be done it's the use of sprinkler systems in a building.'

Too often, though, sprinklers are only a recommendation, not a legal requirement, as Peter Shilton of the Avon fire brigade found when he examined the wreckage of Leo's Supermarket. It 'was not fitted with sprinklers. It was not legally required to be fitted with sprinklers, although the fire brigade had recommended that when the fire certificate was issued back in 1973. If sprinklers had been installed in this supermarket I'm fairly confident in my own mind that a flash-over would not have occurred, the sprinklers would have had the effect of cooling the fire and preventing a build-up of combustible gases, which were enabled to heat and ignite in the form of a flash. Undoubtedly sprinklers would have avoided that build-up, the flashover would not have occurred and maybe Fleur [Lombard, the first female firefighter to die on duty in Britain] would still be with us today.'

The sort of tragedy that can result from a lack of properly enforced building regulations was vividly illustrated in Boston in 1942. Just after 10 p.m. on Saturday 28 November the Coconut Grove's reputation as, figuratively, Boston's hottest night spot and one of the most famous in the USA came to a bitter end. The club was packed beyond its legal capacity with over a thousand people. 'The place was enormously crowded', remembers Dr Dreyfus, a survivor. 'It took them close to half an hour to seat us on one of the tiers. We were seated at a table that was about the size of a dime, and there were eight chairs around it. It was the same throughout the whole nightclub. We sat there and sat there, waited and waited, and nobody came around, nobody offered to get anything for the people there.'

As he remembers the fire, 'We were sitting at this little table and nobody was paying attention to us and suddenly I heard somebody shout, "fire", so I stood up and turned around to see where the voice came from and what I saw was a wall of flames coming along the ceiling somewhere around seventy miles an hour and it was rushing towards us. I put my hands over my eyes, luckily, I think, and the next thing I knew I woke up on the floor of the Boston City Hospital. It was all incredibly quick. There couldn't have been more than about thirty seconds between my seeing the flames and actually passing out,' says Dreyfus.

To this day no one is really sure of the fire's origins, although we do know that a small fire broke out in an artificial palm tree in the Melody Lounge in the building's basement. Even Casey Grant, who has made a close study of the fire and its consequences, admits that 'there's been a lot of speculation towards a match that was lit by a busboy who was attempting to screw a lightbulb back in. There was also concern about the wiring in the facility, which was done by an unlicensed electrician. There were a number of other questions regarding how the fire started but this is academic. What really is the important issue is why it spread as fast as it did and why so many people were trapped.'

As befits its name, the Coconut Grove, says Antony Marra, then a busboy at the nightclub, 'had artificial trees that burned just like if you took a match and lit a piece of paper, that's how fast everything burned in there. You would think they were real but they were all phoney and they burned very fast.' By the time George Graney, a firefighter, arrived with his fire truck the damage had been done. He and a policeman 'started then to remove the bodies. They weren't burned but they were drooped in the chairs and the tables, which weren't burnt. Many of them died from the toxic gases from the fire, which will more than just knock you out. If you get exposed long enough you're gone. The only thing I noticed was the smudges on the nose and the lips and the sensations that run through your mind. You take a person and you feel the back of the neck to see if there was a pulse there or any signs of life but these bodies just drooped.'

In the end firefighters discovered some 200 bodies piled up in the basement. A further 100 were found behind a door in the Broadway Lounge – later investigation revealed that this door opened inwards. The final death toll was put at 492 (this includes a survivor whose wife had died in the blaze and who committed suicide a few weeks later), the worst single fire tragedy in recent American history.

It did not help the survivors that in 1942 treatment, both medical and psychological, was still rather primitive. When Dr Dreyfus recovered consciousness one of the young doctors 'informed me in a very blatant manner that my wife had died, no other explanation at all. I couldn't see at that time, I was literally blind for at least eight weeks and one day they came in and I hard them talking about my right ear and they said, "Well, that's gangrenous, we'll have to cut it off", and they took a knife and cut it off, no anaesthesia, no anything at all. Fortunately in a sense gangrene kills the nerves too so you don't really feel that much. My body was burned somewhere around 37 per cent, all the skin was off my face and my hair was gone, my scalp was burned, my hands, my back was burned, and a whole bunch of other places where I have scars. It was very difficult because as you're recovering from something like this then you begin to hurt and they were very concerned about making addicts out of people so they gave us very little medication. My sense of smell was gone. The only thing I could smell and the only thing I could taste was burnt flesh and that went on for a long, long time, years actually.' Even today, fifty-five years later, after a long career as a doctor, and a happy second marriage, he still does not have 'a single day without pain: 'I always have a pain somewhere and it's not just one. Usually it's two or three or four or five pains but you get used to it, you learn how to deal with it.'

'When you look at how this fire spread,' says Casey Grant, 'it's best to go back and start at the point of origin – forget how it started. Once it got its headway in the corner of the basement, in the Melody Lounge, it took a minute or two then flashed with amazing speed and it turned that single stairway from the Melody Lounge into a chimney. People were attempting to get out at that time but as the fire travelled up the stairway, into the foyer and into the main lobby of the Coconut Grove, it flashed again with amazing speed and people who were there that evening described it as a fireball or as a flamethrower, just coming along as a ball of gas and a ball of flame. It reached every corner of the building in a very, very short time.

'Several things are intriguing about this fire, and among them is the fact that it burned and the flames travelled so fast, much faster than can be readily explained. Why did this fire burn the way it did? Today that's still somewhat of a mystery but it's one that we look at very closely.' When the Fire Department's official investigation was published, it estimated that, feeding on the club's highly combustible decorations, the fire took between two and four minutes to spread some forty feet across the Melody Lounge to the only public stairway out of the building. It flashed up the stairway, through the first-floor foyer and main entrance and into the Broadway Lounge, the club's main dining room. It was less than five minutes from the first sighting of the flames in the basement to the entrance of the fire in the Broadway Lounge – a distance of some 225 feet. By that time all the exits were useless.

As Casey Grant points out, the exits were responsible for a number of deaths. Not only were there simply not enough but 'all of them had something functionally wrong with them. Although the door at the top of the stairway from the Melody Lounge was equipped with panic hardware and swung outward, it was locked and bolted shut. A large number of people were found after the fire at that location. Further, in the foyer, the revolving door, which did have a door on one side of it but which was locked, was serving as an exit and that quickly became jammed. Close to two hundred people were found behind that revolving door.

'The swinging doors in the main dining hall out on to Shawmut Street were locked early on in the fire. They were unlocked by a waiter but the delay caused people to back up at that point. Quite a few people did escape through those doors but quite a few did not, and again a large number of bodies were at that location. The passageway that went from the main dining hall into the new Broadway Lounge was serving as a means of egress in the sense of people trying to get away from the fire. That passageway had steps and it was difficult to negotiate with corners. It was very narrow – again, a large number of people ended up blocking that and died at that point. Finally the primary door out of that Broadway Lounge swung inwards and that quickly became blocked and again a large number of people were found at that location.'

The investigation revealed numerous shortcomings in the building regulations, and the building itself did not even conform to what rudimentary regulations there were. 'At the time of the fire,' says Casey Grant, 'the NFPA life-safety code was known as the building-exits code and it clearly addressed the different features that would have been required in a facility like the Coconut Grove to ensure that everyone was able to get out. When you look at the particulars of what the building-exit code required you see that you would never be able to have a revolving door without swinging doors of the same capacity immediately next to it. You would never allow inward-swinging doors. Doors would be required to have panic hardware – a bar across the door that would allow people pushing on it, as a crowd in panic would do, to have it open. You would never ever have locked doors. You would have multiple paths of egress from any one part of the facility. None of this existed in the Coconut Grove so we see numerous things that would have been required if a document like the life safety code was used.' Unfortunately, as Grant points out, 'In the United States we have the situation where every jurisdiction has to enact their own regulations and in many places those kind of model documents weren't in place or they had rules that evolved over the years that weren't adequate and the Coconut

Grove was allowed to have the design that it did.'

For its part the NFPA claimed that in Boston the building laws were 'chaotic' and that the tragedy was the result of a 'gross violation of several fundamental principles of fire safety, which have been demonstrated by years of experience in other fires and which should be known to everybody'. It wasn't, says Grant, that the 'enforcement applied was inadequate. There shouldn't have been the designs that were in this building at the time and they shouldn't have been acceptable, and yet they met the regulations at the time. The city of Boston, Commonwealth of Massachusetts, were able to look back and realize we needed better rules and regulations.' Previously they seem to have been designed largely to satisfy the city's politically influential building contractors.

The following year building codes throughout the US were revamped. The most notable advances were in the area of exits, combustible materials and emergency lighting. All public areas were to have two separate and remote escape routes. The number of exits into these routes would be determined by the number of people expected to be in the building.

Today, says Grant, 'when you go to any community that has good fire-safety design you'll see that there is an adequate number of exits for the people that you expect in the building, you see all those exits working properly, they all swing in the right direction, they all have the proper panic hardware and other accessories that allow people to get out very quickly, they're not locked. If you see revolving doors you see swinging doors on each side of the same capacity to handle people because the revolving door is not considered an exit door.'

The list of what investigators, in every type of disaster, lugubriously refer to as 'tombstone technology' – their equivalent of shutting the stable door after the horse has bolted – is seemingly endless. Typical was the reaction to the terrible fire at Our Lady of the Angels School in Chicago (see Chapter 12). What came out, says David Cowen, was new legislation 'that mandated that schools be built of non-combustible materials, that adequate early-warning systems be installed to notify the Fire Department and the occupants of the burning building at the earliest stage, that sprinkler systems be installed, that stairways be enclosed, just a whole slew of legislation that has been adopted by not only the city of Chicago but by municipalities across the country and throughout the world so in that regard – I'm often reluctant to say this – the people who died in the fire did not die in vain. I think that's a little too romantic but I guess there is some truth to it. Their sacrifice resulted in some positive change in the world.'

The same applied, says John Keating, a professor of social psychology in Las Vegas 'when they had the MGM and the Hilton fire back to back at a place where the tourist industry is the be-all and end-all. They get very serious about building safe buildings. For instance, the new MGM building is apparently the safest of any hotel in the world. When you have a fire and you see the trauma it causes people and the economic downturn a city experiences in response, you make sure those codes are rigidly employed and that the buildings are very carefully monitored. Unfortunately that's not always the case and we still have buildings that would be problematic if there was an extensive fire.'

In Britain the Offices, Shops and Railway Premises Act was passed in 1973 after a fire at Henderson's store in Liverpool, which cost the lives of eleven people. Like so many fires at that time, it was due to the deterioration of the relatively primitive and badly protected systems fitted in the early days of electricity. (The Henderson fire was worsened by the fact that a number of doors that should have been kept shut had been left open.)

Many of the lessons that should have been learned from the Henderson fire were ignored in the construction of the Summerlands leisure centre in Douglas on the Isle of Man where fifty people died in a blaze in 1976. Summerlands was the island's response to the growing fashion for holidays in the sun, well south of Britain. In the words of a local architect, J. Philipps Lomas,* it would provide 'the leisure life of the twenty-first century' not in a building but through a whole complex with 'a weatherproof enveloping structure'. To give people the illusion of sunshine in the open air, the roof and the vast wall structures had to be 'transparent, glare-free and tinted to sunny-day colours. So, virtually from the earliest design and planning stages, the use of acrylic "glazing" was a firm choice for the roof and the south wall.'

Unfortunately the project illustrated every possible mistake, making the result a guaranteed disaster for the Douglas town council and the Isle of Man government were clients, promoters and the regulatory authority. The architects they employed had no experience in major schemes, let alone one

*Quoted in *The Thin Red Line*, Stephen Barlay, Hutchinson Benham, London, 1976.

as revolutionary as Summerland, and found it difficult to co-ordinate the project. Even worse, the work was parcelled out among a host of sub-contractors.

The final element in creating a disaster was that the island, always stubbornly independent, had not passed legislation similar to that enacted in Britain two years earlier, the long-overdue Fire Precautions Act. Four days before the disaster the local fire chief told the *Isle of Man Courier* that the existing fire precautions at Summerlands were 'pitiful'. He was duly reprimanded for his outspokenness. But, of course, he was right: the building was a true fire trap. The acrylic cladding had been applied in a manner guaranteed to make it inflammable, and the material used in the inner lining of the wall, itself made of Galbestos, a British Steel product with good fire-resisting qualities, was changed – probably by a builder anxious to save some money – from relatively safe plasterboard to combustible fibreboard, leaving that most dangerous of traps: an inaccessible void in which a fire could gather momentum – even though the widely publicized fire at Henderson's had been fanned by just such a cavity construction.

Summerlands' operators, Trust House Forte, were also at fault because it was not clear who in the management chain was responsible for fire precautions, fire drills and other safety measures. As a result, the staff did not know how to operate the elaborate alarm system and, although they behaved heroically, they made many mistakes – the worst by the house manager, who turned off the main electricity supply, thinking that this was the right thing to do, thus leaving in the dark those trying to escape through clouds of smoke. Miraculously 'only' fifty perished of the 3,000 people inside the building at the time. If the fire had occurred two hours later the numbers of the dead would have been far higher as 5,000 would have been caught up in the conflagration.

The resulting inquiry produced a relatively mild report. Nevertheless, it did establish that the fire had been started by children in a kiosk close to the shopping centre, which had collapsed against the side of the building. The Galbestos, though fire-resistant, had not been designed to resist the transmission of heat by conduction and the fibreboard naturally ignited. The children were caught and admitted that they had started the fire with matches. For causing 'wilful and unlawful damage' to the lock of the kiosk they were fined £3 each, and ordered to pay 33p compensation and 15p costs.

In many modern buildings, sheer size and the complexity of the equipment they contain can militate against efficient fire prevention. In the report of the US National Commission on Fire Prevention and Control, its authors point to additional dangers built into modern high-rise buildings, such as permanently sealed windows, which allow air-conditioning ducts to serve as efficient spreaders of fire. The other major problem is the way in which a fire can spread with disconcerting rapidity through a large complex. 'Fire starting in one area and tragedy occurring in another is a not uncommon phenomenon in these major-loss fires,' says Tom Klem, fire protection engineer. In large buildings 'products of combustion move from areas that lack adequate sprinklers, and move, gaining momentum, through corridors and penetration points within buildings. Occupants are trapped then killed by the toxicity of the smoke. People succumb to the products of combustion remote from the fire so it's a phenomenon that's really associated with large structures that have high densities of people and complex mixed-occupancy types, like storage and hotel occupancies, public-assembly occupancies, like a casino, and so forth.'

The *locus classicus* of the terrible effects of fire in a complex high-rise building was the MGM Grand Hotel and casino, with capacity for 5,000 guests, in Las Vegas in 1980 when 85 guests were killed. It was the worst hotel fire in recent American history. It was, says Tom Klem, 'a classic example of fire in a remote location that resulted in fatalities occurring far from it. The majority of the fatalities in the MGM were within the high-rise portion of the building. Only ten were at the same level of occupancy as the fire origin. The smoke, products of combustion and toxic gases moved through the high-rise areas by a number of routes. One was the heating, ventilation and air-conditioning system, the other was through seismic joints.'

The fire had started, says Klem, in 'an unoccupied restaurant area in the ground floor of the building. It grew in a concealed space within the busboy station and spread in the restaurant area. Before it was detected it had moved out into the casino area, on the ground floor of the MGM Grand. The rapid spread of fire was a result of its going undetected within an unprotected area of the building, growing in magnitude until it breached the outer perimeter of the restaurant, when it moved with a great amount of energy through the casino. As the smoke and toxic gases moved through the casino, there was a natural tendency for those very buoyant gases to move up.' This they duly did, fuelled, as independent fire analyst Carl Duncan says, not only with 'many tons of combustible wood resin mastic at the ceiling but there was also the pilot ignition of concentrations of gaming machines, furnishings in satellite lounges and service areas. There was also ceiling failure – metal grids suspending gypsum board began to buckle and warp – and heat and smoke.'

'The mass of smoke', adds Ralph Dinsman, battalion chief, 'had to go some place and it went up smokeless stairwells at the end of the hotel. They had seismic joints that were about a foot wide, covered with a type of aluminium that had melted from the heat of the fire. The smoke spread rapidly up to the twenty-sixth floor of the hotel and, because roof vents were not open, the smoke started to mushroom, spreading out, going across the floor and then started working its way down. Later on in the day, when we were able to asses what exactly had happened, we found out that from about the twenty-second to the twenty-sixth floors were where most of the fire deaths had occurred.'

When it got to the guest rooms, says Klem, 'it distributed smoke within the corridor areas of the high-rise portion. Because of the make-up of the heating, ventilation and air-conditioning systems within the rooms, it actually circulated smoke from the public corridor into the rooms, so there was a circulating medium from the exit access corridor into the room by way of the heating, ventilation and air-conditioning system for the individual guest rooms.'

To make matters worse, says Klem, a lot of people died in the stairwells because of modifications in the building, though these were within the law at the time: 'One of the stairways ran horizontally across the level of the casino and it was protected with gypsum material as opposed to masonry material. That was code complying. However, the gypsum barrier had been penetrated by workers gaining access to utilities and the like. As a result there was a large hole in the horizontal portion of that exit way right over the casino area. As the fire grew and spread in the casino the buoyant products of combustion moved vertically and into that hole and the circulation of those toxic gases throughout the stairway occurred as a result of that. So it was a very tragic result in an area where it really startled investigators. We perceive stairways as a protected environment – once you reach the stairway you're going to safely exit the building – but that wasn't the case in the MGM.'

There was a dramatic contrast between the MGM tragedy and a fire at the Hilton nearby only three months later when 'only' ten people died. It was a spectacular blaze, an arson, Klem says, 'that occurred in the upper level of the high-rise, hotel portion of the building. The fire was severe and spectacular: it traversed the exterior of the building, leaping from floor to floor and mostly involving the elevator-lobby area of the hotel as opposed to spreading significantly in the interior of the building. The other aspect that makes the Las Vegas Hilton different from the MGM and others is just the way that the part of the building that burned is arranged. The fire occurred in the guest-room area. It is designed in separate compartments, and each individual room is compartmented from the next adjacent room. That type of arrangement and the location, coupled with the location of the fire, led to a more favourable outcome, tragic none the less, but it wasn't the high loss of life that we experienced in the MGM Grand.'

The dramatic difference that design can make was vividly illustrated by the contrast between the following two fires. One, in 1974, killed 220 people when it destroyed a newly built office block in São Paulo, Brazil. In the other, in Boston, ten years later, not a single life was lost when guests were safely evacuated from a thirty-eight-storey hotel and the leisure complex. At São Paulo the fire started on the eleventh floor and spread upwards, rapidly engulfing the top fourteen floors. Some of the 650 people trapped on the upper floors tried to escape by way of the exterior balconies. They were filmed by TV crews below as they jumped to certain death rather than wait to be burned to death. The Jolema Building was only a year old but lack of fire-safety legislation meant that it had been built with highly inflammable materials, and without a fire escape.

In the Boston blaze the Westin Hotel had a fire-resistant structure and was stuffed with a galaxy of fire-protection features. These included pressurized exit stairways, automatic fire-detection systems with heat and smoke sensors, automatic alarm systems with voice messages for guests directly linked to the Boston Fire Department, and sprinklers to extinguish any fire before it could spread. Those who perished forty years earlier in the fire at the Coconut Grove had not died in vain.

# 5
# Fatal Bits and Deadly Bobs

The uniqueness of the fire was that we had such devastation and loss of life by something as simple as a few rags discarded improperly with some linseed oil.

Bob Buckley, ex-fire investigator, Philadelphia fire marshal's office

Today – and not before time – building materials are divided into two types: combustible and non-combustible. They are tested in a furnace heated to 750°C, and if they give off inflammable gases, or burst into flames for ten seconds or more, or even if the temperature of the furnace rises by more than 50°C they are categorized as combustible. Although they can be made more fire-resistant or, at least, 'retardant' – able to delay the effect of flames – they can never change category because they provide fuel for fire and give off smoke and toxic gases. However, even a naturally non-combustible material like steel may collapse if it is subjected to intense heat.

Of course, experts like Casey Grant, assistant vice-president of the NFPA, will point out that 'we have standards that allow us to make sure that the materials that are used are not going to burn or give off poisonous gases as may have occurred that evening in the Coconut Grove fire.' The Boston Fire Department even employs a chemist, whose duties include testing the inflammability of decorations in public places. Yet materials are still used that cause innumerable blazes.

The potential killers include natural products like paper and softwood, which catch fire at relatively low temperatures, and 'thermosetting' plastics, like polystyrene and the polyurethane used in rigid foam, which ignite at a higher temperature but which then burn rapidly. The worst liquid dangers are petrol and other hydrocarbons – even diesel ignites under pressure without a spark – liquid solvents and alcoholic spirits. The most dangerous materials are self-heating, like soft coal and oils of vegetable origins – cod-liver oil, tung oil, cottonseed oil, and linseed oil.

In the 1960s and 1970s innumerable tragedies involved burning polystyrene tiles, plastics and the foam products used in furniture, not only because of their inflammability but also because they gave off highly toxic gases. In 1970 the crinkly plastic substance covering the walls of a huge nightclub in Grenoble, eastern France, was set alight by a burning match and gave off dense clouds of smoke. A few moments later fire erupted and the choking smoke killed nearly a hundred and fifty young dancers. It was the fumes of burning paint and plastic from floors and window-frames that killed over two hundred office workers in São Paulo. In August 1970 some seats in the BOAC terminal at Kennedy Airport, New York, caught fire. They were made of layers of plastic and foam rubber covered with plastic upholstery. The fire, it was said, leaped 'like a red panther in a small cage'. A 1973 British government report referred to 'restaurants decorated with natural materials, basements full of old newspapers; burning silk and wool release deadly quantities of carbon monoxide and cyanide gas – and these and many other natural materials ignite at lower temperatures than many synthetics do'.

Obviously buildings in which inflammable materials are used are also at serious risk of fire – garages, paraffin stockists, fried-fish shops, distilleries and bonded warehouses, laundries and dry cleaners.

Just as dangerous are careless workers whose activities and the processes and products they use are not always properly checked and supervised. Airport terminals are usually thought of as safe: precautions are strictly observed outside because of the effects of gases from aviation fuel. However, in 1996 at Düsseldorf airport a blaze killed seventeen people and injured a further sixty-two. It had been caused by welding work on the expansion plates in a roadway outside the terminal. They were located above the false ceiling of a florist's shop at the bottom level of the building and the ceiling space was filled with highly inflammable polystyrene foam ignited by sparks from the welder's equipment.

Opposite City Hall in the heart of Philadelphia, awaiting demolition, are the towering remains of a thirty-eight-storey granite and steel behemoth, One Meridian Plaza. At one time it housed some of the city's best-known law firms, communication and insurance companies, but in February 1991 fire destroyed the top nine floors at a cost of several hundred million dollars and the lives of three firefighters.

When George Yaeger, a battalion chief with the Philadelphia Fire Brigade, arrived on the scene he saw 'heavy smoke coming from one window at approximately the twenty-second floor. I felt that it was a fire in an office and that although it would be manpower-intensive to take under control, as most high-rise fires are, we could successfully extinguish it'. In such cases they would normally 'take breathing apparatus, hose line, forcible entry equipment to the floor below the fire, stretch the line from the standpipe and attack the fire in an interior aggressive manner. At the floor level or at the base level we would pressurize a standpipe with our pumpers and pump water up the standpipe to the twenty-first floor.'

It soon started to go wrong. While Yaeger was talking to another chief firefighter 'There was a loud pop and all the electricity went. . . . From that point things deteriorated rapidly. We were now faced with very little water pressure on the twenty-second floor and no elevators because the electricity was out. The building engineer who came to the first floor tried to get the emergency generators going. He was not successful. The auxiliary fire pumps we use to supply water were lost too so we had no electricity to the building, no elevators, very little water pressure and the fire was growing rapidly.

'Every firefighter had to walk at least twenty-one floors to get to where they could attack the fire, which created time constraints and a need for much more manpower'. Of course it involved an appalling strain on the fire-fighters, too: they were wearing 'bunker pants, boots, a coat and a helmet, which together weigh about fifty pounds,' as well as 'self-contained breath-ing apparatus, which weighs twenty-five pounds, so they're starting out with seventy-five pounds on their back before they add any equipment, such as hose, nozzles, axes, forcible entry tools, anything of that nature. They could be carrying up to two hundred pounds at any one time.' Not surprisingly, 'It took from twenty minutes to a half-hour for each crew to get from the first to the twenty-second floor and it's a laborious climb. It required some rest periods which was frustrating the firefighters because they're used to getting on the scene and immediately going into service.' When they finally got there, connected their hoses to the standpipes in the emergency towers and got only a trickle of water, they had to evacuate in fear of their lives. The pressure-reducing valves in the standpipes had been set incorrectly. In desperation the firefighters connected hoses to the water hydrants at street level and lugged the hoses up the full twenty-two floors. But by that time the fire had taken such a hold that they couldn't control it.

Yaeger and his colleagues faced further problems in that every floor in the three towers was laid out differently and conditions grew worse as the fire spread. By then many more firefighters were on the scene but Yaeger realized their efforts were in vain when the fire extended to the twenty-third floor. 'In my experience, it was rare in a high-rise for a fire to extend from floor to floor. Once the building became unsafe to operate in the commissioner felt that he would not risk the firefighters. At about eleven o'clock that evening three firefighters got lost above the fire, were overcome by carbon monoxide and died. An engineer told the commissioner that this building was structurally unsafe to operate in, that the potential existed for it to collapse, so his decision was to pull everybody [around 300 firefighters] out. They knew that there was a sprinkler system on the thirtieth floor and, the hope was that it would contain the fire, which it did. At least that was ultimately successful.'

When John Malooly and his colleagues in the ATF's national response team arrived on the scene, his first task was to try to establish which was the room where the fire had first been seen. 'When we arrived on the twenty-second floor,' says Steve Avato, special agent, ATF, 'we observed a scene of total devastation. It appeared that all combustible materials on the floor were burned, the units that hold in dropped ceilings were hanging loose, there were large steel girders that were twisted and bent. We were concerned about the structural stability of the entire floor. We had some people who came in and looked at the steel beams to make sure they wouldn't fall on our heads, that's the kind of damage that was done in here: there was enough heat produced to twist steel beams.'

As a result, says Bob Buckley, 'The condition of the building made the investigation very difficult. All the equipment we used during our investigation had to be taken up manually. It took almost a full day to get up to the floor where we needed to do the investigation. Then it always seems like there's something else you need so you're running up and down. Eventually some elevators were running in the building next door where they got a connection so that we could use them.'

He goes on, 'Because they had to abandon firefighting operations at some point, there was not much for us to follow. Unfortunately normal investigative techniques, such as following burn patterns and analysing the fuel load, were not much use because the building had burned significantly and was allowed to burn without any firefighting activities on it. Those normal burn patterns were damaged so we had to rely heavily on eye-witness accounts.'

Fortunately two eye-witnesses confirmed where the fire had started. The investigators had been told by the firefighters, says Buckley, 'that an employee had been at his desk when the alarm went off on the twenty-second floor. He went up the elevator to check it out – he didn't call the Fire Department. He opened the elevator and was hit by dense black smoke, which drove him to his knees. He had to call down to the desk to get one of the other workers to bring him down manually so that was the first indication that we had some type of fire on the twenty-second floor.'

'We knew early on,' says John Malooly, 'that the fire began on the twenty-second floor and spread upwards. We were fortunate in having a video taken by a passer-by of the early stage of the fire. We did a freeze-frame on the video and counted the windows from the side of the building until we reached the one that had flames venting out of it. We were also able to notice on the video that flames vented out of that office, then skipped an office, then vented out of another office window. We logged all those fail-ures of the window glass and needed to explain why that middle office was spared. Why did it not become involved until later on when it was closer to the origin of the fire?

'We dig down like an archaeologist. We layer through the fire debris. After the fire structural members fall so we reverse the process and go through layer by layer. In this case, we wanted to see the position of the hinges in the lockwork on all of the office doors, which doors had been left open, which doors were closed. By doing that we found out why the fire skipped that particular office – the door had been closed but the adjacent one was open so the fire skipped past one office to the next. Once we got into the room of origin we layered down through the debris and found painters' cans, turps, things they were using to finish some of the wood in some of the offices.'

'At a fire scene,' says Steve Avato, 'we sift through the debris layer by layer to locate any potential sources of ignition for the fire. We're looking for electrical components that might have generated heat that ignited the first fuel; we're looking for any heat-producing devices, such as portable heaters. In this case we were aware that there were chemicals that may have been able to produce spontaneous heating so we sifted through the debris looking for any causes of a fire. We found a pile of rags that had a lot of surface burning on them but the centre of the pile was not heavily burned. This indicated to us that the fire may have started in or around this pile of rags. There were also some very solid rags, which rags typically are not, that led us to believe that spontaneous ignition could have occurred at this scene.

'As the investigation progressed we believed that spontaneous ignition was most likely in this case but because three firefighters had lost their lives, there had been a lot of damage to the building and disruption to the people of the city, we wanted to make absolutely sure that that was so. We spent a lot of time eliminating all other possible sources of ignition.'

The Meridian building was relatively old but had never previously caught fire. They soon found an anomaly, says Malooly: 'There was some finishing going on inside the office suites. . . . We asked the tenants what was going on and they said these people were in the process of refinishing some of the solid oak, walnut and cherry panelling and they had done this every few years. It was very high-quality wood so they were using a natural oil finish on it, which was principally linseed oil. We began to layer down in that building and we found remains of the cans and the containers and tools that were there and one area of some deep-seated burning was discovered. Down in that area we found the remains of some coarse-woven cotton rags, which had been used to apply the linseed oil and which were very deeply charred.'

They had already interviewed the contractors and, says Bob Buckley, had 'found that they were finishing off some cherrywood panelling using boiled linseed oil, paint-thinners and some other stains. Our concerns would be how they disposed of their materials.' They had told the investigators, says Avato, 'that all of the rags that they had used during that day were placed in a paper bag and taken down to a trash receptacle on the ground level. We did find in the trash receptacle a bag with some rags in it but we believe that the rags that were found were not all of the rags that were used that day. That led us to believe that some rags were used and had been soaked with linseed oil that did not make it into the trash dumpster on the ground floor, that some may have been stored in the room when the fire occurred.' So, says Malooly, 'we gathered up the rags and we placed them in evidence containers and forwarded them to our laboratory. The laboratory found linseed oil on those rags so we knew that the rags were not removed from that office as we had been told.'

Linseed oil can combust spontaneously when the oil reacts with air, so it is perfectly safe when it is stored in an air-tight container. If the heat building up is not dissipated as fast as it is being generated then it quickly reaches a temperature at which it ignites. In the case of fabric soaked in linseed oil, this could take a mere couple of hours. Bob Buckley explains that, 'Linseed oil creates an oxidizing process whereby it makes its own heat and its own oxygen and if allowed to continue without dissipation of the heat it'll generate temperatures significant enough to ignite the material that it's on. Generally what'll happen, you'll have cotton rags and there'll be piles of 'em bundled up and the heat's not allowed to escape from the core of the rags so it just builds up and builds up and ignites. The conditions have to be just right and they were in this room. It was a closed room, the ventilation system was shut off for the weekend and we were told by one of the engineers that it gets very hot in the room. There's no movement of air. Linseed oil generates very little smoke in the early stages of a fire and when it does ignite it generally starts slowly.'

Avato was not alone in experiencing 'a tremendous feeling of satisfaction when we finally found a cause for the fire that we could all accept. Everyone in the investigative team was determined to locate it – any time a firefighter or anyone loses their life there's a certain emotional bond between us, a need to know what happened. In this case when we found the cause, it was a little disappointing because it's hard to believe that something as simple as a pile of oily rags could have cost the lives of three firefighters and destroyed a thirty-eight-storey skyscraper.'

Five years later the cause of the fire in the Garley Building in Hong Kong was just as extraordinary. At around 5 p.m. on 20 November 1996 people stopped in the street to look up at this fifteen-storey office building. Smoke was billowing from the top three floors. By 6 p.m. when the main TV news went on the air the whole of Hong Kong watched as choking, burning people waved towels to attract the attention of firefighters below. Fifty died, trapped in the building, leaving an enduring image of doom in their helpless bid for rescue.

Many survivors have terrifying stories to tell. Kenny Leung, one of Hong Kong's best-known dentists, had two surgeries in the building. He was lucky: he didn't have any patients that afternoon. If he had, he says, 'it would have delayed at least by two to three minutes my escape from the office. I might not have made it at all. When I look back I think it's really a miracle that I was able to get out of my office. If I only had one patient, then I would have to ask the patient to sit up, rinse their mouth out, walk out and try to be as graceful as a dentist should. I don't think we would ever be able to make it with that time delay. The second thing is I was very thankful that there was no panic when we walked down the stairs. . . . The smoke was blocked off by a brick wall and did not spread to the staircase. If there had been smoke then people would have panicked and there would have been a stampede.'

His traumatic experience began when 'Somebody working in the opposite office knocked at the door and came in and told my dental assistant that there was smoke in the corridor. My assistant called me. I walked out of my office and I went out to the corridor to take a look. I saw a few people already gathering in the corridor. They were having a party, they were still kidding and joking with each other, saying what kind of a welder is there producing that much smoke. In the corridor the air was actually clear. We smelt smoke but we didn't actually see any smoke in the corridor. Then as we looked to the far end of the corridor, which is the lift lobby, we saw dark, heavy black smoke and I didn't think that's caused by welding alone so I figured there was a fire. I reopened my glass door, I told my assistant to call the police immediately.

'Then I looked back at the corridor and there was nothing, but I had the strange feeling that this black smoke was growing in density and I had a feeling that it's coming towards us and I felt a little heat so I told the other guys to get out of the building for a while to let the firefighters do their job. The other guys all said OK, OK. We went back to our respective offices and everybody started evacuating. I wetted a stack of towels and I threw some to my assistant. She left, I went back to my office and grabbed my wallet and my keys. When I was back in the corridor I could not see anything, like, five or six feet away – the dense smoke had already filled the corridor and I couldn't breathe without covering my face with a wet towel. I forgot about locking the door and it's fortunate that the fire escape was just right next to my door.'

The fire was localized. For the first three or four flights of stairs down towards the ground, 'There was still smoke but as I walked down the smoke became clearer and after three or four flights of stairs I didn't even have to cover my face.' Once he reached the pavement he looked back, but 'There was nothing except a little smoke coming out from the elevator end of the building, a little dark smoke. . . . After around five minutes we heard glass breaking and we looked up and saw people from the thirteenth floor breaking the office windows and then we saw the smoke and the fire inside the office building and then they started climbing out of the windows, standing on the air-conditioners. They were shouting for help and trying to escape.' Unfortunately, the fire engines were on the other side of the building.

The rescue effort involved two helicopters and first on the scene was piloted by Mike Ellis, who remained the very model of the cool British pilot, even though it was the first time he had ever tried to rescue people from a crowded city centre. The fire, he says laconically, 'was easily visible because there was a huge pall of black smoke you could see from twenty miles away'. In the narrow streets of central Hong Kong 'all the buildings tend to be either high-rise or very tightly packed together. We had to position ourselves down between the high-rises and fly down the street on about the fourteenth-floor level, between the buildings on the other side of the road and the building that was on fire, to be able to get down close enough to the building roof to see if there was anybody on top of the roof that we could take away.'

When he arrived he saw 'the flames coming out of the two upper floors. It was seriously on fire – I mean, very intense flames coming ten or twelve feet out of the building. You couldn't even see the top wall of the building because of the flames and smoke.' To put it mildly, visibility was impaired: 'We've got obstructions on all the other rooftops, aerials, unmarked hazards, things hanging off roofs and actually a lot of debris started to come out of the fire itself, flowing up, flowing down, blowing everywhere. We had two thousand pounds of fuel on board so didn't want to get too close to the fire. We were moving forward very slowly, visibility's down to probably five to ten yards and our blades are sticking out thirty feet in front of us so we're moving forward very tentatively. These little rotor blades are moving around three hundred and something miles an hour, so we're very careful about that. We've got debris coming from the roof we don't want to get in the engine intakes, and we don't want anything going through the rotors. The primary concern is the winchman on the back. We want him in a safe position and to make sure he's safe all the way through the operation. We don't want to lose him.'

The rescue was hazardous. 'We were going to have to establish a stable hover over the corner of the roof,' says Ellis, 'long enough and low enough to be able to get the winchman on the end of the winch wire on to the corner of the roof. He had to pick up these four guys one at a time as quickly as possible and hoick 'em into the aircraft.' There were two abortive efforts, but 'The third time we came in we stayed right over the very corner of the building, we stayed very low, kept a steady position and got the winching carried out as quick as we could. . . . We were getting fire-hose water spraying over the front of the building, actually coming across our windscreen, making life more interesting. . . . And we'd got about the second guy up and my co-pilot, Johnny Lee, turned round and said could we come up a little bit because the flames were coming past my wheel on my side and it was getting a bit warm. Obviously we had to try and maintain the hover as long as we could to pick the guys off – and the paint didn't melt so we were OK.'

Even when he had rescued four of those stranded his difficulties were not over. 'The only way out was by the way we'd come in, back down the street, which we had to do backwards and in smoke and all the fumes.' But he managed, with the help of his winchman and the pilot of the other helicopter who had joined the rescue effort.

There were some amazing stories of survival – and terrifying ones of mortality. Dr Leung saw one boy 'standing outside the window of the first office unit on the fifteenth floor. Everybody was shouting to him, "Do not jump, do not jump", and obviously we saw fire coming at his back. Then he was holding the window panel and I think that's kind of hot. . . . After a minute or two he just jumped from the fifteenth floor. But he was lucky because the building had a podium on the third floor with a tin roof and he landed on it. His body bounced up and down a couple of times, and then he was just flat there and my legs were shaking. I said, "Man, this boy is dead", but after a few minutes we saw his hand and his leg move a little bit. Then debris started to fall from the higher floors, the air-conditioners, broken glass. So I thought even if he wasn't dead by jumping, he'd be dead when he was hit by an air-conditioner. Miraculously, after a few minutes, I saw him crawl to the side of the building', where he was rescued by the firefighters.

The rescuers could not save everybody. One tenant, a friend of Leung's, had opened a window. 'The firefighters saw him and they were raising the ladder up to the fourteenth floor but at that time an elderly man had climbed out of the window and he was sitting on an air-conditioner. The ladder was swinging back and forth as they tried to decide who to save first.' Because the older man was already outside the building they rescued him. By the time they returned for Leung's friend 'he was out of sight. Later on they found his body burned, lying just by the window.'

Because the drama had been played out live on TV as it happened before the majority of the population and because the death toll had been high, there was intense popular and political pressure to find out what had happened. The Hong Kong government's forensic laboratory was asked to mount an investigation. 'It was a very special fire,' said investigator Bobbie Cheung. 'We used a number of investigators and worked for a whole week to get to know what really happened in the building.'

From the outside – an impression reinforced by the TV pictures – the fire looked as though it had been confined to the top of the building. But on their first day on site the investigators got a terrific shock: the *bottom* three storeys of the Garley Building were the most severely affected by the fire. In fact, as Cheung says, 'When I walked through the building I had quite a different picture from what I'd seen. From the TV pictures the upper floors of the building suffered severe burning but when I walked in the building I found that the lower floors had also been damaged severely while the floors in between the top and the bottom were only mildly affected by the smoke. So the fire could have been started on the top and spread down or vice versa.'

But had the fire started at the top, the damage there would have been the greater. As it was, both top and bottom were equally affected so, says Cheung, it was 'more likely for the fire to have been started on the lower floor and subsequently spread up to the top floors. We concentrated our examination on the lower floor first. After looking for burning patterns and directional signs we found that the fire was more severe around the area close to the lift lobby on the lower floors so we started detailed excavation to get more information about the materials there.'

On his first day on site Cheung had been surprised to find that 'all the lift doors of this building had been taken out and the cars removed so that when we looked down from the top floor through the lift shaft to the bottom it was like an empty tube. There was renovation work taking place in the lift shaft so they had had bamboo scaffoldings built inside it.' The empty shaft had been the ideal means for the heat and fumes from the fire to move up – it had acted like an extremely effective chimney.

The effect, says Cheung, was like that of an old-fashioned Bunsen burner. Once you open the hole at the bottom, he says, 'fresh air goes in, mixes with the gases and burns more severely at the top. In the Garley Building, at the bottom of the lift shaft the ground-floor opening was like the air-hole of a Bunsen burner. The fresh air goes through from the ground floor and the fire on the lower floor generates a lot of smoke and hot fumes, much like the gas in a Bunsen burner. They mix in the lift shaft then cause very severe burning at the top floors.'

By the end of the second day the investigators knew, from the patterns of burning they traced on the bottom three floors, that they needed to get into the lift lobby area. When they cut through the distorted metal fire partition into the lift lobby on the second floor they found charred carpeting and builders' materials, which indicated the source of a smouldering fire. But how could a fire that started on the second floor miss out the rest of the building and jump to the top three floors?

The answer lay in the fact that a new lift was being installed. The lift cars had been removed, and the doors on all floors had been replaced with temporary partitions. On the top three floors these partitions had been removed because the workmen installing the lifts had wanted more light. Once the smouldering fire on the second floor had taken hold, hot gases and smoke went up the lift shaft as if up a giant chimney. They blew out along the corridors of the thirteenth, fourteenth and fifteenth floors. This left the question of how the fire started. The investigators discovered that welding had taken place in the lift shaft that afternoon. But that had been on the twelfth floor: how could it have sparked off the disaster?

The only way to find out was by reconstruction of the events. On the seventh day of the investigation they set up in a parallel lift shaft and videoed the results. Within five minutes the flux (the material used in welding) fell ten floors down the lift shaft, out into the lift lobby on the second floor and ignited the materials placed there by the investigators. The fire had started when welding flux hit cardboard packaging material and was intensified when it spread to hollow, highly flammable bamboo scaffolding poles. But the interior space in bamboo poles is not continuous: it is separated into cells at the joints. As the heat of the fire increased, the air inside these sealed spaces expanded until the bamboo exploded, sending fragments of burning wood flying around. Just as in One Meridian Plaza, the old-fashioned materials that were used to repair a modern building caused disaster.

# 6
# Stop the Home Fires Burning

We can engineer death out of buildings through sprinklers in exactly the same way as it's been done in the United States in people's homes and it can be done with relatively low-cost systems, very very effective systems which are highly reliable, they don't spoil your home and they will save your life.

Malcolm Saunders, deputy fire chief, West Yorkshire Fire Brigade

Despite the publicity given to major fires in large public buildings, deaths from fires in the home are overwhelmingly more frequent. In the early 1970s domestic fires accounted for seven out of ten of all fires, and nearly nine-tenths of all loss of life from fire in the USA; in Britain four out of every five deaths from fire were domestic. However, for journalists, who so often reflect the interest of the public, and indeed many fire investigators, domestic fires are small, sad events that make the local news for a day and are then forgotten.

There are innumerable causes of domestic fires. In 1970 an American commission* listed examples of dangerous products, ranging from electric blankets to gas heaters, dryers and hotplates. All are now safer than they were, but still may be worn or badly connected. And there remain serious dangers: the death rates from fire in caravans or mobile homes is nearly three times that in static accommodation. And accommodation for the elderly, where special facilities are required because of the vulnerability of the occupants and their inability to move quickly, is often hazardous.

However, a host of new regulations and a generally increased awareness of the dangers of fire have ensured that the numbers killed in domestic blazes has fallen steadily over the past quarter of a century in both Britain

*The National Commission on Product Safety.

and the USA. The biggest single contribution to fire safety in British homes was made by Bob Graham, MBE. When he retired in 1993 he was head of fire investigation for one of Britain's most densely populated conurbations, Greater Manchester. It was in the early 1970s, though, when he was in charge of a station at Kidderminster in the West Midlands that he had an experience that changed the face of Britain's homes. It was the result of a fire in a factory that made cheap upholstered furniture. The fireman had tackled the blaze in the conventional way but 'something went wrong,' Graham says. 'Two firefighters were trapped. In the course of the rescue, I and another colleague were injured and two firemen were lost. After I came out of hospital the people who manufactured upholstered furniture and foam came down and talked to us about foam. I was aware of most of the hazards of foam, having seen it in domestic environments. Their view was that it was no more dangerous than conventional materials, wood, paper, that kind of material. They said it had a similar heat release as wood, and on that basis it was no greater hazard than wood. I couldn't believe my ears when they were telling me this. I made the point that it releases the calorific value of heat in seconds rather than in the hours that a piece of wood would take and I came to the conclusion that either they weren't aware of the consequences of fire involving their products or they were deliberately obscuring us.'

Graham had already started collating information on this type of fire but the industry's response so irritated him that he started 'to look more closely at the problem and do some research on it, find out what work had been done by the Fire Research Station and other organizations. Gradually a picture built up that prophesied that fire deaths were going to go through the roof if we didn't do something about it.

'In the late sixties, early seventies, there was a marked change in the nature of domestic fires. Prior to that a lot of fires were what we call kitchen and contents. If the furniture was involved in a living room, it was traditional furniture, upholstered with horsehair and those kind of materials, and it burned slowly. What we were getting was fires that completely destroyed everything in the room.' Typically, 'People go to bed at night and someone has dropped a cigarette end on a couch or an easy chair and it falls down the gap between the side of the seat and the cushion. They go to bed, the fire incubates, starts to smoulder and after about an hour or two hours it starts flaming. Within three minutes of that occurring the whole room is ablaze.

People are not awakened by the smoke. It's curtain rails burning through and dropping curtains that spread the fire, cords on pictures burning through and dropping pictures which break and wake them up. That's their first indication of the fire and by then it's too late.

'The problem with these fires was that once furniture upholstered with polyurethane foam was involved it increased the likelihood that people were not going to survive.

'Once the fire starts you have this flaming combustion and it starts spreading very rapidly over the cover and over the foam itself. It starts to motor, goes faster and faster and faster. The flames get above the back of the settee and then, before you know it, they're hitting the ceiling and heating everything else up and everything else starts to vaporize. All of a sudden you get flash-over and everything is burning all at once. Now, the smoke that comes off burning polyurethane foam includes hydrogen chloride, hydrogen cyanide, carbon monoxide. When you say hydrogen cyanide people think, "Oh, that must be dangerous", and it is, of course, but the fact is that this stuff produces so much carbon monoxide in such a short time that these other highly toxic gases are irrelevant. Carbon monoxide is the killer, and it's the rate at which it is produced that is important rather than the strength of the toxicity.

'The smoke is at a very high temperature – commonly you get temperatures at ceiling level of up to 1000° centigrade. The first time you encounter that – if you were upstairs and you started to come downstairs and tried to go through that smoke – it's the irritants that get to you. You've taken a breath and you can't open your eyes because as soon as you do they water. You take another breath and the irritants hit the back of your throat. You retch and take a deep breath – it's a natural involuntary reaction – of these very toxic fumes. That disorientates you, puts you down on the floor, and while you're incapacitated there the toxicity takes over. You succumb to the effects of carbon monoxide particularly.'

Firefighters were experiencing the results in the changing nature of domestic fires hundreds of times a week up and down the UK. But, as Graham points out, 'These were what people regarded as small incidents, one or two fatalities. It wasn't a major factory fire or a hotel fire where five or ten people were dying and that seemed to be tolerable to society.' Moreover, and crucially, until Graham came along, firefighters and scientists didn't really work together. 'When I was in Lancashire in the late sixties,' says Graham, 'I had a particular incident where there was dreadful damage and fatalities in a house fire – the police thought it was a crime and they brought in a forensic scientist to investigate. He saw the damage and said immediately that it was a petrol bomb attack. I can understand him saying that because he'd not seen a furniture fire before. When we explained the way the fire develops and showed him further evidence of this he agreed with us that this wasn't petrol or a petrol bomb', but that it was the result of polyurethane foam.

In 1977 Graham returned to Manchester, and in May 1979, 'we had a fire at the Woolworths store on Piccadilly. It was "persons reported", that means persons are known to be in the building and are trapped. The fire brigade sent a major response to this incident and I was mobilized to it as well. We had a very serious fire on the second floor and, the smoke that was coming out of the building was unbelievable for just a furniture department to be involved. The fire didn't spread to any other floors – although the smoke spread through the building – it was contained, because the building was well constructed.'

Although the fire was confined to a single floor, and the brigade arrived within a few minutes of the alarm being given, 10 of the 500 people in the building were killed, 53 were admitted to hospital and 26 had to be rescued by firefighters. The blaze was so fierce that within a few minutes even fire-fighters wearing breathing apparatus couldn't go near it.

Bob Graham took his chance. Although he was still relatively junior in the force he convinced his chief to allow him to conduct the brigade's investigation into the fire. The speed with which it had spread had taken most fire investigators by surprise. The ten people who had died had been right next to an exit when the fire was spotted, yet had failed to get out.

Bob Graham thought he knew why. 'I put a team together of people who arrived at the building when the rescues were being carried out, as well as an officer who had been involved in issuing the fire certificate – he knew everything about the premises. I told the officers to start collecting information while I looked at the main area of burning. I knew the building well. My job at that time was fire-safety legislation and I knew the building was due to be issued with a fire certificate. Everything had been done: there were adequate exits, there was a fire-alarm system, there were fire extinguishers. I found it hard to believe that people were unable to escape from a building that met all the standards. When I went on the floor it reminded me of a house fire. It was a huge floor area but the damage was the same. It was widespread everywhere, down to the floor. There was heat and smoke damage over the whole of the floor area and that isn't the kind of thing that you get in lots of buildings when you have a fire. We knew something had happened there that had stopped these people escaping. They had known about the fire, they had seen it on the floor when it started. It was puzzling: what had happened that stopped these people escaping?

'The area where the fire started was very severely damaged. There was little there that was recognizable. Right down to floor level everything was blackened and charred, not necessarily burnt because the mixture of gases in that kind of situation is so rich that it can't burn, it only burns when it comes out of the windows, but the heat level was such that plastic was burned right down to floor level.'

As he observed, the damage was similar to that in a house fire but he couldn't 'get a grip on the fact that this had happened in a large department store. At that time we'd been having bomb warnings, incendiary devices and so on in most major cities in the UK, and Manchester was no exception so that was a consideration at the start. But from the evidence of the survivors we realized that this was something different. It had gone so fast that people just couldn't move away from it quickly enough. The problem was that, with a fire like this, which produces a great deal of smoke and toxic fumes very quickly, which travels across the room at high level, when it hits a vertical obstruction – the enclosing walls and doors – it comes down to the floor, so people moving away from the fire from a clear area quite close to it came upon an area that was impenetrable between them and the exits and that's where we found them.'

'You couldn't really see more than a foot in front of you. It was just choking,' says Steve Wood, a survivor. 'People were putting their hands or cardigans over their mouths to try to breathe, people were yelling – I can't really describe it any more, it was just a situation I don't want to be in again. I don't think many people would want to be in that position either.

'I was surprised how quickly it happened. Obviously I'd been to bonfires, when you're in an open space and the smoke and the heat normally goes upwards, but in a building the smoke just went through every nook and cranny to find its way out.'

Significantly the fire had started in a department containing furniture covered in polythene wrappers, with sheets of Kraft paper round the settees and the armchairs. 'We started to get information that there had been upholstered furniture in there,' says Bob Graham, 'and, more importantly, that it had been stored in the area where there was a great deal of fire damage. Things started to add up in my mind. With my background of upholstered furniture fires, I realized that if we were to convince people that this was a real problem, this was one opportunity that mustn't be missed. But they might not believe me because I'd been saying these things for many years so I called upon the services of the Fire Research Station (FRS) to come up and help me to investigate the effects of the fire.'

The replica fire was set up in the FRS' testing facility in the old airship hangar at Cardington. 'With the help of Woolworths, says Stan Ames, fire consultant, 'we were able to identify every piece of furniture that was in the area where the fire started and obtain those from the manufacturers. We had them shipped down to the FRS and we reconstructed with the help of Woolworths' managers the stacking of furniture in the area where we found the fire started. Then we lit it.'

The FRS was well prepared to help Graham prove his theories. 'When the Woolworths fire occurred,' says Ames, 'we had already been studying the inflammability of furniture for some years and had begun to realize that the things people like Bob Graham were saying were true. We'd been doing research for the government's own purchasing authorities and we were looking at the way furniture ignited, the way fire spread in it, how fast heat was released, how big the fire was from this sort of furniture, and we were beginning to realize that the levels of toxic gases released were much higher than we'd expected. Most of the furniture we studied burned quite readily but one type worried us particularly. If the furniture was covered in a fabric that melted like polypropylene and contained padding material that melted like unprotected polyurethane foam, this produced a pool of burning liquid underneath the furniture and this liquid accelerated the speed and growth rate of the furniture fire, making it far more dangerous. This type of furniture was very popular: it was very cheap, there was an awful lot of it about, and this was the very type of furniture that was present in the Woolworths fire.

'We had the advantage of already being involved in a research programme but we also had at our disposal this giant fire laboratory, the largest in the world at that time, and it was possible for us to do a complete reconstruction of the fire that occurred on that day in Manchester. We assembled a complete stack of furniture exactly the way it was in the Woolworths store and ignited the fire, which grew very quickly to the point where flames were reaching the ceiling. From this point onwards the fire growth was more to do with how much smoke was being produced because we could see that large amounts were moving fast under the ceiling of our test rig. In reality this would have moved right across the ceiling of the Woolworths store and people there would perhaps not have been aware of the hazard that this represented. Maybe this explains why a lot of people there didn't react as quickly as they were expected to.'

'When it was being set up,' says Bob Graham, 'some people who were in the fire business, fire engineers, fire consultants and so on, were sceptical, and I remember one of them coming to me and saying, "Well, what do you think is going to happen here?" and I said, "You won't be stood here in two minutes," and I think there was some disbelief there that this thing could go so quickly. The fire was lit and the rest is history. It went so quickly that the TV cameramen recording the scene had to leave their cameras. They had to warn everyone to evacuate because the fire started to motor.

'We were already aware that this type of furniture would burn rapidly,' says Stan Ames, 'but we'd never seen it behave in this way because we'd never studied a stack of it before. Put simply, if you've got five or six armchairs, put them side by side on the floor and ignite one at the end, it takes some time for the fire to grow from one chair to another. If you take the same six armchairs and stack them vertically, then ignite the bottom chair, it will ignite the ones above it very, very quickly. Sure enough, this was what we found when we did our experiment.

'Perhaps the most important thing we found was how much smoke and toxic gases were produced. This stack of furniture produced a very high level of dense, toxic black smoke and we found at the peak of the fire that it was emerging from our test rig at the rate of 1,700 cubic metres a minute, which would have filled that area of the store very quickly. The other thing was that we were measuring the toxic gases in that smoke. The carbon monoxide, which is a narcotic gas, in that smoke would have put people to sleep in two or three breaths, which probably explains why so many people were caught before they were able to escape.'

The test was a milestone in the investigation of fire. The size of the Cardington facility 'allowed us to do something which would otherwise be very dangerous. We were able to study this fire under closely controlled laboratory conditions with detailed scientific measurements, but perhaps, above all, we filmed our experiment and the videos of this fire have been sent all over the world. Since that time they've been used widely to teach firefighters and fire engineers their trade.'

At the same time, 'We learnt a lot about the behaviour of furniture in fire, we learnt a lot about the movement of smoke and the toxicity of smoke in these fires. We learnt that the characteristics of this fire were the same as those that fire brigades were reporting in domestic fires. The injuries and deaths in this fire were similar to the injuries and deaths seen in domestic fires, both linked to the fire behaviour of this type of furniture.'

Graham was pleased that he'd proved his point 'but we weren't there yet. This was a fire in a department store. The configuration of the furniture, being stacked, had a lot to do with the way the fire spread and developed so quickly but this same situation was happening in hundreds of homes every week and that was the thing I wanted to change.'

The Woolworths fire enabled Graham to bring the problem of polyurethane foam to the attention of the public, the authorities and, finally, the politicians. At the time of the fire, TV crews had been on the scene. The image of salesgirls trapped behind barred shop windows as fire engulfed them was already imprinted on the publics' consciousness. At the inquest, Graham says, 'There was a lot more media coverage about the killer in our homes, those kind of phrases, but the only thing you could do was ban hazardous materials. Polyurethane foam is a hazardous material, it is easily ignited and it burns so quickly that you could be in the same room and you wouldn't have time to escape before you were overcome. We worked through the eighties, particularly in the Greater Manchester brigade, and we gathered a lot of information on domestic fatalities. Every incident that occurred throughout the country was passed on to us and we focused on the contribution in those fires of polyurethane foam upholstered furniture. It was quite clear that half of those fires became killer fires because of the involvement of the upholstered furniture. It didn't matter how they started. We made sure these were brought to the attention of the media. Some newspapers took up this as a campaign to get the furniture banned.'

In one particularly dreadful fire, early on Christmas morning 1984, nine people died. 'Christmas wrapping paper caught fire and set fire to a polyurethane foam cushion. Two people were sleeping in the room and they escaped, but none of the nine people upstairs managed to get out. That

A fire impossible to douse, when a million tyres blazed in the heart of Philadelphia by the side of Interstate 95, one of the busiest highways in the United States. (KYW TV3)

**Above:** The grim result of the 'tunnel effect' which devastated King's Cross underground station in London in September 1987. (Rex Features)

**Left:** Summerlands, the leisure complex in the UK's Isle of Man doomed by bad design and construction. (Noel Howarth)

**Above and below:** It only took a single cigarette end to do enormous damage to Hampton Court Palace, one of Britain's most famous historic buildings. (Rex Features)

A heap of packaging and some bamboo scaffolding poles were responsible for the deaths of 50 victims in a fire in the Garley Building in Hong Kong in 1996. (Bobby Yip/ Popperfoto/Reuters)

Terror in paradise: the Du Pont Plaza building beside the sea in San Juan, Puerto Rico, scene of a conflagration which triggered a major advance in the science of fire investigation. (Jose Fernadez/Rex Features)

Once a fire has really taken hold, all the firemen can do is prevent it spreading. (Mitch Kezar/Tony Stone)

The grim, grey aftermath of a fire: realistic – but not for real. The lifelike staged fire in a Chicago warehouse created for the US television drama *ER*.
(Michael Springer/Gamma Liaison/Frank Spooner)

focused even more public attention on the problem and I was asked by the coroner to investigate the way this was being dealt with in other countries. I did a study in the States, where they used a fire-resistant foam, that was published in 1985 but still there was this resistance to any controls at all on the foam. The government set up a working party, I managed to get on that but mainly it was industry representatives who were there and they produced a code which they said would solve the problem. In my view, it didn't do anything to solve the problem. In fact there was a view that it would make the problem worse. The fire brigade decided to come out with the politicians and with the media to argue the case for controls on polyurethane foam.'

The years 1986 and 1987 were bad for foam fires. Finally, 'Working with the opposition – I think it was Tony Blair who was the shadow trade spokesman – we managed to convince the government that they should do something about it. In 1988, regulations came in which banned ignitable covers and the use of untreated polyurethane foam. This year we have seen the difference that can make. We've had a look at the fire statistics. Untreated polyurethane foam fires [on furniture manufactured before the ban was imposed] numbered something like 1,500, accounted for 40 deaths and something like 500 injuries. [By contrast] combustion-modified polyurethane foam accounted for three fires, one death and five injuries. As long as we stick with it and don't allow a watering-down of standards, the benefits to society are the biggest on fire safety that I think there will ever be.'

However, perhaps we should also ensure that the sort of positive safety features normal in commercial and industrial premises are fitted in every home. In the late 1970s this approach was taken in Scottsdale, a suburb of Phoenix, Arizona, when residents were asked to vote on the future of their fire service: did they want a full-time force or a part-time force, which would be much cheaper? If they wanted the part-time force they had to agree to support a new city building regulation, requiring all new private dwellings to be built with automatic sprinkler systems. They voted for the sprinklers, and since then there has not been a single fatality in any of the houses fitted with them. This would not surprise Ed Comeau, chief fire investigator, NFPA: 'In ninety-nine per cent of fires, if a sprinkler system had been there the fire would not have occurred.'

Jim Ford, Phoenix's assistant chief, makes the decision to install sprinklers sound so simple that it is astonishing that Scottsdale's example has not been followed. 'We looked at putting sprinklers in homes because that's where most people spend most of their time. They should feel and be safest in their homes. But if you look at the statistics that's where we lose more lives. Four out of five people who die in fires die in their homes. We decided to look at how we could do that better and, given the effectiveness of sprinklers and how well they've worked in commercial structures and the testing that we had done, we found out that they could be used in homes. We did that in 1985 and it's worked extremely well for us since then.'

According to Ford, the sprinkler systems have been 'much more effective than we even dreamt of. We have had about 150 fires in sprinkler buildings. The value of those structures was over $600 million and we had a fire loss that equalled less than the price of one house. We've found an average fire in a non-sprinklered house causes about $17,000 loss. We've found that houses that had the sprinkler systems had an average of about $2,000 loss.'

More importantly, 'We know for a fact in this past ten years that we saved at least eight people's lives.' The most recent example was 'a young man who was in a residential structure that, typically is not protected with sprinklers. He had come home, had gone to sleep and his room-mate had decided that he wanted to take his identity. He poured gasoline on this young man while he was sleeping and set him on fire in an attempt to kill him and burn the house down – he thought that nobody would be able to tell. The sprinkler system went off, put that fire out, saved that young man's life and saved the structure.'

Ford's laconic account of the Scottsdale experience plays down the revolutionary change in approach it embodies. As he says, 'Typically, what happens is "legislation by catastrophe". That's when a major event occurs in a city or a community and you go to the policy-makers and you pass legislation to address the event that's already occurred. It makes everybody feel real good but in all honesty it's too late. What we've tried to do here in Scottsdale is pass legislation ahead of time to address those problems, especially the life-safety problems, and it's worked extremely well for us.' He makes a contrast with the authorities in Gothenburg who have come to Scottsdale to learn how to impose a universal sprinkler system – after the Stardust disco fire described in Chapter 3, which cost sixty-four lives.

Ford believes that, generally, 'people are reluctant to change and I think that specifically applies to built-in protection and sprinkler systems. They're not familiar with it, they don't understand it, so why change things that have worked this way for years? I think a good analogy is seat-belts in cars. When these first came out nobody wanted seat-belts, now people wouldn't even think about buying a car without seat-belts or air-bags because they understand the benefits. I think the same thing will occur with built-in residential sprinklers. In the future people are going to understand the type of protection they are receiving for very little cost and very little problems of getting those in homes.'

And in Britain? In a reversal of their usual role, firefighters from the West Yorkshire Brigade set fire to a block of flats in Cleckheaton as part of a programme aimed at reducing fire deaths. Malcolm Saunders, the brigade's deputy fire chief, wants the installation of sprinklers to be accepted as standard practice in new buildings. He believes that 'We can engineer failures out of fire. What we've learned from investigation of fires is that the only way to survive them is for effective suppression to take place before the fire service arrives. We're very efficient at putting fires out but unfortunately those people who die do so before the fire service comes. Sprinkler systems can stop that.'

But his approach is under attack from some sections of the fire-investigation community. According to them, it is how people behave when caught in a fire that should dictate the design of our buildings.

# III

# THE HAND OF MAN

# 7
# People Who Hate People

Arson is a crime like any other crime, it is different in a lot of ways and some people have compared it to homicide. But in my opinion it's sometimes worse than that because if I shoot a gun at a person I intend to kill I might hit them or I may hit innocent victims. But there's limited damage that a bullet can do, it will stop just by the laws of physics at some point, fire will continue to burn as long as it has fuel and oxygen or until the fire department can stop it and sometimes they can't stop the fire. So unlike a bullet that will eventually stop, fire will just continue to burn and destroy whatever is in its path.

Steve Avato, special agent, ATF

When fires are caused deliberately, as they often are, the fire investigator turns into a detective. The transition itself is not difficult, but the discovery of the criminal and bringing him (he is usually male) to justice, presents extraordinary problems, greater, perhaps, than in any other type of crime.

Bob Bell, senior forensic scientist, Forensic Science Service, echoes the views of most experts on the subject: 'Out of all crime probably arson is one of the easiest to commit. However, it's probably one of the most difficult to investigate and solve. All the criminal needs to do is to set fire to something and with a bit of luck it will burn away as he desires. The problem with investigating is that most of the material has been destroyed.'

Lieutenant Renney Pelszynski, assistant fire marshal, Philadelphia Fire Department, agrees: 'It's a clandestine crime. It's rarely done with witnesses to observe what goes on so we're hard pressed as investigators to come up with motives and with witness information that'll actually place the torch into an individual's hands. That's the most difficult part of our job and our success rate both in this city and probably on a national level would be somewhere between 16 and 20 per cent in solving who did it and getting a prosecution into a courtroom.'

Because of the difficulty of defining, let alone solving, the crime of arson, no one really knows the proportion of fires that are started deliberately. We know that it is high and that it is systematically underestimated in fire reports, which tend to be cautious in their conclusions. The report on the tragic Woolworths fire considers the possibility that it was caused either by an electrical fault which was inherently improbable, or by 'direct ignition', which could mean a carelessly dropped cigarette end or an arsonist, but ends, 'There is insufficient evidence for a more specific conclusion.'

So high is the likely proportion of arsonic fires that in his standard work on fire investigation John D. de Haan* states flatly that *all* fires should be 'treated as a potential arson scene (from the standpoint of security, preservation and evidence) until clear proof of natural or accidental cause is discovered'. The figures seem to back him up. In 1994 the NFPA estimated that at least a seventh of all fires in the United States was the result of arson, accounting for at least 86,000 fires and 550 deaths. The previous year, its British equivalent, the FPA, reported that nearly half all the large fires in Great Britain were almost or certainly ignited deliberately.

One of the many problems facing investigators is that those who set fire to buildings and their motives are so varied. In the United States the National Center for the Analysis of Violent Crime (NCAVC) has classified motive in six neat categories: profit, vandalism, excitement, revenge, crime concealment, extremism. Of course, any of the six can be sub-divided: 'extremism', for instance, may include fire-bombing by religious fanatics (of mosques, synagogues, churches, abortion clinics) or the burning of the shops of supposedly exploitative retailers (Korean in Los Angeles, Chinese in Indonesia, etc.).

In his book *Arson Investigation*† Robert Carter gives a flavour of the variety of motives involved:

Builders hoping for work, conspiracies involving the assured and the insurance assessors, business competitors, watchmen and security guards looking to justify their positions, students to avoid attending classes or as a form of protest, strikers damaging their employers or

*Kirk's Fire Investigation*, fourth edition, Brady/Prentice Hall, Englewood Cliffs, New Jersey, 1997.
†Collier Macmillan, London, 1978.

warning off blacklegs, or racial conflicts in which blacks set fire to shops owned by perceived enemies or whites encouraging new black neighbours to move.

Carter illustrates his categories with a variety of cases. A fire department officer had been passed over in favour of a much older man. His son set fires in the hope that the older man would not be able to cope with the workload and would have to retire, thus leaving room for the boy's father. In another example, a boy was ridiculed for 'being effeminate'. His pride and joy was a flash new car, and he burned the cars belonging to his rivals – or even cars he had seen them admiring in a showroom.

De Haan points out that commercial blazes may be started 'by a competitor to gain a market advantage or by agents of organized crime for purposes of extortion, insurance fraud, protection rackets, or intimidation. Labour problems are a frequent cause, particularly in the building trades. During construction, buildings are especially susceptible to fires set by union or anti-union agitators.' Fraud is another obvious motive, ranging from the over-insurance of a warehouse full of sub-standard goods to trying to get more than the value of a decrepit old car. And, yes, there have been cases of fires in warehouses still full of Christmas cards in January.

As we see in Chapter 10, the arsonist may be an insider. The largest anti-arson operation in Britain in recent years, Operation Nero, resulted in the detection of an arson-fraud conspiracy that had been responsible for the destruction of a dozen or more buildings. The guilty party was Peter Scott, a professional loss assessor, who advises victims of fires on how and what to claim.

The one thing arsonists appear to have in common is that they don't feel guilt. This is clear from observations made by Dian Williams.* She is a 'criminal profiler', and defines her work as 'largely a matter of combining and compiling information and developing a picture of the probable individual who might have done that particular crime'. When she is teaching interrogators she always tells them 'don't spin your wheels looking for guilt, people who are fire-setters generally set their fires on purpose. If they were going to feel guilty about it they wouldn't have set it to begin with. I think it's human nature for those of us who feel badly about doing something

*President and CEO of the Centre for Arson Research in Philadelphia.

wrong that we want to believe that somebody who would deliberately do a bad thing would feel horrible about it afterwards. In fact, for the most part, fire-setters don't feel anything at all.'

But for all the reasons for this particularly anti-social crime, the people involved seem more like criminals than anecdotal evidence might suggest. An analysis by the NCAVC based on interviewing eighty-three convicted serial arsonists showed that they had set an average of over thirty fires and had started their life of arsonic crime at the age of fifteen. They were over-whelmingly male, more than four-fifths were white and they were not well educated. Over half had already been in juvenile detention and nearly half had mental health problems. A staggering 87 per cent had previously been arrested for some other form of felony. They didn't have to travel far to indulge their criminal fancies: over two-thirds of their crimes had been committed within two miles of their homes. And their motives? Two-fifths were looking for revenge – mostly against society – and another 30 per cent were looking for excitement.

Dian Williams divides fire-setters into seven or eight sub-types: 'Some are accidental setters, generally little kids who might find a pack of matches and be fooling around and set something on fire, become very frightened and never do it again. We interviewed a nine-year-old who said that he happened to find a pack of matches and he took it into an abandoned barn and he dropped them in a bale of hay and the barn burned down. Well, when we interviewed him he was convinced of two things: one, that he was going to go to prison for the rest of his life; and, two, that he was going to be sent to his room until he was thirty-five, which to him sounded like it was going to be for ever. What we knew after the interview was that it was truly an accident. He was in that barn because he had stolen one of his dad's cigarettes and was trying out smoking. He thought he heard somebody coming and he dropped the lit cigarette and then it caught fire to the barn. That kid would never set another fire in a million years. He made a mistake and he was very frightened and very upset, so we don't have to worry about him.

'A second kind of fire-setter is the kind of person who is hearing voices under some delusion in which they feel that it is incumbent upon them to set a fire to save the world or to prevent some kind of catastrophe or to protect people against the devil. Once they have been treated for their psychosis their fire-setting behaviour goes away.

'A third type is incorrectly called pyromania, which is a very antiquated expression left over from the French Revolution.'

It's not only antiquated, it's vague and unreliable. As de Haan points out: 'There is no consensus even among psychiatrists and psychologists as to what constitutes pyromania, and that it may not even be considered a disorder.' It may also be part of a pattern of even more serious crimes. David Berkowitz, the 'Son of Sam' killer who terrorized Boston in the late 1970s, claimed credit for setting over two thousand fires during a three-year period. But his feelings of loneliness and social inadequacy found their most complete expression in murder.

Arson is often connected with other crimes as an attempt to destroy evidence or disguise a murder as an accident. In 1979 Bedford School was badly damaged by fire when burglars set light to the late Victorian Great Hall in an attempt to cover their tracks. More gruesomely, arson helps murderers: it is difficult to detect whether their victims were killed by fire or whether fire concealed murder. Fire boils blood, causing haemorrhages and broken bones. The skull is particularly vulnerable, which can create the false impression that an individual who died in a fire was intentionally killed by a blow to the head. In November 1930 one Alfred Arthur Rouse tried to disguise a murder by setting fire to a car with the corpse inside. Although the identity of his victim was never established, Rouse was hanged for the crime.

Dian Williams talks of the serial arsonist 'who sets fires in or around an anniversary date of importance to them. They set maybe thirty-five, forty fires in a row, day after day, for about two weeks, to all kinds of different locations. They don't stay and watch them, they just set and go, and when that anniversary time is done they're finished setting fires for the year.

'Delinquent fire-setters usually start in early adolescence. That's the one group that will set fires with one another. They also have other criminal behaviour and if they're not setting fires they're doing something else. They are dangerous because, unfortunately, they don't really care about the consequences of their acts. Even when they say they're sorry, what they're generally really sorry about is being caught. When they grow up, most of them will move away from fire-setting but many graduate to other criminal behaviours and a lot end up in the adult criminal-justice system.

'Thrill-seeker fire-setters are another of the sub-types. They start doing dangerous things early in life. They jump from the roof of a garage or run down the side of a cliff or dive from high places. As they become teenagers, if they drive motorcycles, they ride them without helmets and they go through red lights. They love danger, they love the edge and their fire-setting behaviour is a game. At the Centre we entitle the game "I'm smarter than you and therefore you'll never catch me, and if you catch me I will never admit it". They are the group that like to use timing devices. They like to play the investigative game, leaving few clues and seeing if anyone else will find those clues. Unfortunately in fire departments you may well have a thrill-seeker or two around. They're attracted to the danger in their work and if there isn't enough danger or enough excitement it's not unheard-of for them to become fire-setters. They never care about loss of life or the probability of injury or destruction. It doesn't matter to them at all.

'The last two types are deliberate fire-setters. They have a very early age of onset, almost always under the age of five. If their fire-setting behaviour is not interrupted by the onset of puberty they will set fires for their whole lives. The oldest person we've ever interviewed who as a fire-setter was seventy and he was still an active fire-setter, had never been caught, had set hundreds of fires of all different sizes and used fire as the way to cope with anything that made him feel uncomfortable.'

Fire-raising might be used as a way out of a psychological predicament. As de Haan points out, 'Juveniles may set fires as a solitary activity to express anger or seek attention. They may set them as a cry for help because of an abusive (physical, sexual or psychological) family situation or in deliberate response to a threatened departure (as in divorce) or unwanted arrival (new step-parent, child or lover).' Indeed, juvenile fire-setting, for a wide variety of reasons, has been greatly underestimated: nearly 7 per cent of those arrested for arson in the United States in 1994 were under the age of ten.

'Most fire-setters,' says Williams, 'are male, about 6 per cent female, and they are generally found in this last sub-type, called revenge setters, another dangerous sub-type. They are also the most likely group of all to be fire bombers. Their behaviour begins around eleven and twelve, they have a get-even mentality and often grow up in families that teach getting even as a value, and also that if someone harms you, there is no such thing as an accident so you can never forgive but you must always get even. They're very dangerous. Of all the sub-types there is only one group that returns to the scene to look at the damage that they've done and that is the thrill-seeker. They will, because it's important to them that they set big fires and they're the only group that cares about that. If the fire is not big enough they will reset it.'

When the investigators feel that a fire was or might have been set deliberately they are looking for what de Haan calls 'abnormal fire behaviour or unusual fire conditions, incendiary devices, fuels, or simply things that are out of place for some reason'. These should 'sound an alarm in the examiner's brain that warns of criminal activity'.

Steve Avato of the ATF says, 'There's always excitement at the beginning of a fire investigation when you arrive at a scene and you want to get in to see what the scene has to tell you, to see what clues are left and where those will take you, to see how and where the fire started. If you're going to have any success with fire investigations, usually there's a good break in the beginning of the case that really helps out. That'll often come within the first twelve to twenty-four hours.'

Investigation is hampered because, as de Haan points out,

> The official investigation of arsons seems to fall into a gap between the area of responsibility of the fire department and that of the police. Fire department personnel may consider their primary responsibility to be public safety through extinguishment of the fire. They may consider the warrants, interviews, and fieldwork necessary for a criminal investigation to be outside their normal authority and more appropriate for the police. Police investigators, on the other hand, realize that the physical evidence of the crime is beyond their training and may refuse to deal with it. As a result of this schism, valuable evidence of one type or another is often overlooked.

They just have to grit their teeth and embark on a detailed look at all the evidence. As Chris Porreca, manager, arson and explosives program points out, 'In today's prosecution of arson cases the systematic approach is the only way that you can do it. Technology has shown us that we have to be very systematic in everything we do, not just in fire investigation but any other law-enforcement activity. In fire investigation you have to use an approach that, when you get into court, you can relate to the jury and to the defence, and that was systematic and methodical in everything. You must show that you took into consideration everything that was in that scene that you were examining and that you either ruled that out as a possible cause of the fire or you attributed it to being one of the causes of the fire.

Bob Bell, senior forensic scientist, Forensic Science Service is more specific. 'What we're looking for,' he says, 'are those traces which haven't been destroyed. We can look for accelerants, but we can't find a match. I suppose those are the most difficult ones to investigate, where somebody's just used a naked flame – the match either burns away or they take the cigarette lighter with them – but, then, we're not so much looking for what's left but what could have really caused that fire, so we move into a slightly different sphere: how was the fire actually started?'

Clues are not confined to the scene of the crime. In his discussion of commercial frauds de Haan combines internal and external evidence. At the scene there may be 'substitution of cheap, old furniture, equipment, or stock for new; open drawers in file cabinets or desks to encourage destruction of records; remodelling underway with minimal or poor-quality materials; or disabling of fire detection or sprinkler systems'. Away from the scene,

> the background investigation should determine recent changes in ownership, value or tenancy; the state of business – debts, back orders, financing, and mortgages; the nature of any inventory – contents new or old, seasonal, outmoded, obsolete, or unsaleable because of changes in government regulations; and the extent of competition. Contact with the insurance agent and adjuster will reveal the amount of insurance, multiple policies, or recent changes in valuation or coverage.

The most obvious clue in a case of arson is that a fire was started in two separate points. But H. T. Yallop* provides two cases in which impressions were misleading. In one case, at a timber factory, a fire had started (or been started, he wasn't sure) in some wood shavings at one end of the building. The hot gases generated by the fire had travelled the length of the building, emerging through a ventilator near roof level. There were also fires at every point at which there was a skylight, allowing the fire to reach the piles of wood below and giving the – erroneous – impression of arson.

In another case fires started in a bar at the same time near a closed roller shutter and at a seat well away from the shutter. What had happened was that a current of hot air had travelled along a ceiling channel. Once the fire was hot enough to cause a flash-over, the burning particles ignited some curtains, which fell and set fire to material below – yet there were no signs

*Fire Investigation, Alan Clift, Solihull, 1984.

of intense heat in the ceiling channels.

The investigators are not entirely helpless in the face of the arsonists. Bob Ludwig, a fire marshal in Philadelphia, has developed a technique that uses ultraviolet light to reveal traces of hydrocarbons. If the arsonist has stepped in petrol on his way from the scene of the crime Ludwig can find his footprints. In Newcastle upon Tyne, arson had reached such epidemic proportions that a task force was set up. A criminal psychologist, fire investigators and a hydrocarbon dog* joined forces. In the first five months of the project the detection rate went up from one in twenty-five incidents to one in three.

From time to time fires or, rather, series of fires, can assume a different dimension. So it was with the extraordinary fires, mostly in the state of Washington – above all in Seattle – in the late 1980s and early 1990s. They all bore the traces of what became known as high-temperature accelerants. Such fires did not merely burn large industrial premises: the extraordinarily high temperatures they generated melted steel and turned concrete into a glass-like substance. 'It was like science fiction,' says Richard Gehlhausen, then an investigator with the Seattle Fire Department.† 'There's not yet been a building designed that can withstand this fire,' says Dennis Fowler, a consultant and former Seattle firefighter. The mysterious HTA generates such extreme temperatures that water merely feeds the flames, probably because at the temperature of 4000°C or so, characteristic of such blazes, the water splits into its components – hydrogen, which explodes, and oxygen, which fuels the flames.

The HTA blazes were characterized by a speed and suddenness unique among fires, and an intensity never before found in peace-time conditions. The first occurred in 1984 in a carpet warehouse in Seattle. Afterwards Fowler found that a roof truss made of steel had not merely melted: part of it had vaporized and part of the concrete floor had been transformed into a shiny turquoise-coloured glass-like substance. This phenomenon occurs only in temperatures similar to those experienced when a rocket is launched from Cape Canaveral. The idea slowly emerged among researchers that the arsonist was using rocket fuel: not only does it produce the unusually high

---

*Officially known as 'accelerant detection canines', they are trained to sniff out petrol and other accelerants through the stench of a burned-out building.
†Quoted in an article by Erik Larsen in the *Wall Street Journal*, 7 October 1993.

temperatures found in HTA fires, it is designed to burn in an atmosphere free of oxygen.

The Seattle Fire Department asked for help in determining the cause of this and other such blazes, such as the one that occurred in the Blackstock lumber yard in September 1989 in which a firefighter died. A dozen or more institutions became involved, including a number of defence laboratories concerned with the nature of rocket fuel, although, because this is designed to provide thrust for the rocket rather than flames for a fire, it contains too much aluminium to be entirely suitable for an arsonist.

The investigators were aware, as firework manufacturers and rocket scientists had been for decades, that a number of metals, ranging from aluminium to titanium, might serve as high-temperature fuels if they were blended with a suitable oxidizer, like the ammonium perchlorate normally used as a fertilizer, and concentrated their researches in experimenting with such metals. They reached a climax in March 1990 when they set alight an empty store at Puyallup, Washington, with dramatic results.

The investigators had devised a suitable compound and had assembled devices to measure the temperatures reached by the blaze in several locations and in a firefighter's coat. The fire started quietly but after two minutes was transformed by a flash-over. The whole building virtually evaporated. The most awesome evidence came from the measuring device in the firefighter's coat, which was designed to cope with temperatures of up to 1300°F. Within two seconds the temperature had soared to 500°F 'Meaning,' said Dennis Fowler, 'that if you were lying on the ground you'd have two seconds to get out of there.'

The effect of such mixtures was confirmed by the experience of the US Navy during the Gulf War when the USS *Stark* was hit by two Exocet missiles launched by an Iraqi aircraft. The devastation caused by the subsequent fire was attributed to the rocket fuel carried on board the ship – a result confirmed by subsequent tests at the US Navy's research establishment at China Lake in California using ammonium perchlorate and a suitable chemical to bind the fertilizer-like material together.

Although everyone involved agrees as to the nature of the fuel, there is considerable disagreement as to the number of fires caused by HTAs in the years when they were being set. Some investigators believe that there were up to two dozen, scattered around the USA, as far afield as Florida and Illinois. This was largely because the enormous publicity generated by the

HTA fires ensured that every fire department in the USA was on the look-out for fires that might have involved this phenomenon.

But the most exhaustive study of the HTA blazes, carried out by Steve Carman of the ATF,* concluded that most of the fires presented as HTA-related either were not or that sufficient data was lacking to make such a conclusion. He also warned fire investigators to differentiate between HTA fires and ordinary fires where the heat was released with exceptional swiftness. But he also provided a working definition of the fuel used: 'a combustible mix of metal fuel and a solid oxidizer' giving a temperature of 2500°F or above. Such mixtures were found in 'rocket fuel, fireworks mixtures and thermite compositions', which would produce the characteristic mix 'of intense white or nearly white flames often restricted to a relatively small area', similar to those produced by a welding rod or carbon-arc light and often accompanied by 'displays of white-hot pyrotechnic-like sparks'.

Carman examined twenty-five fires, thirteen in Washington state, that had occurred between 1981 and 1991. To me, though this may be unfair, his study bears the mark of efforts by his superiors to downplay the phenomenon, probably because the ATF had limited resources and enough then on its plate to want to avoid the enormous task of solving the HTA mystery. Carman dismissed the much-touted idea that some sort of conspiracy was involved. As he puts it, 'As with any criminal cases, investigators must rely on firm evidence rather than mere supposition before claiming the existence of grand conspiracies, particularly in the media arena.'

In the end he classed only four – all Seattle fires – as 'potential HTA blazes'. These included the first fire to have aroused suspicions of HTA involvement, at the Carpet Exchange. One of the most significant pieces of evidence came from an injured firefighter. Even before the building had exploded in characteristic HTA fashion a firefighter walking through the building felt a 'burning pain in his right foot. The injury involved third-degree molten metal-like burns, although there were reportedly no visible burns to his shoe.' When a shoe and debris from the fire were analysed, they contained traces of nearly a dozen inflammable metals, including aluminium and titanium. The debris was so corrosive that, within a few days, the cans in which it had been packed had rotted away. Similarly, at the fire at the Golden

*High Temperature Accelerants: A Study of HTA Fires Reported in the United States and Canada Between January 1981 and August 1991, BATF, October 1994.

Oz restaurant in Kitsap County the flame temperature was judged to be 4000° – 5000°F and evidence that when water was poured over the flames it simply made matters worse.

Carman's caution – and the keenness of investigators to categorize any unusually hot blaze as due to HTAs – meant that he listed as 'undetermined' the cause of fires like that at the First Christian Church at Bellingham in Washington State, even though cast-iron radiators had melted in the heat, and Carman admits that 'the history of [previous] fires in the building offers a sharp contrast in behaviour compared to this fire'. He also refused to judge fires without enough forensic evidence, which ruled out a number of cases and even a 1987 fire in a lumber yard in Indiana, in which some of the concrete had turned blue-green and witnesses had seen flames of the blue or white hues typical of an HTA blaze.

He also dismissed as 'unlikely' a suspicious fire in June 1991, in a small town in Pennsylvania, a warehouse and factory belonging to the Galaxy Cheese Co. burned down. The insurers were forced to pay out a cool $11million.

But more importantly for other investigators, he dismissed the HTA claims of the most tragic of all such fires, at the Blackstock Lumber Company in Seattle in September 1989, in which a senior firefighter was burned to death. Five years after the blaze, it was found that a 440-volt power line was still connected to the city's power supply, a spark from which almost certainly caused it.

However, local investigators believe that up to seven blazes – all in Seattle or within striking range of the city – were caused by HTAs. They analysed the phenomenon closely, realizing that an HTA was useful for an arsonist anxious to destroy a large building with the minimum of materials. HTAs, in melting the steel and concrete framework, ensure the total destruction of a building more effectively than orthodox blaze-inducers. All the buildings were well insured – and one, the Blackstock Lumber Company, was at the heart of a complicated property-development scheme which required its destruction for completion.

Now comes some fancy detective work. From two separate official sources, Gehlhausen learnt, albeit indirectly, that in the 1980s a Russian Jewish rocket-fuel expert had emigrated to the USA, had worked in a defence company, but had been sacked after building up gambling debts. A further twist to this story, according to Gehlhausen, is that the debts could have been owed to a leading member of the local Mafia for whom the émigré's expertise would have come in handy in the destruction of well-insured business premises. Another, less tenuous link was that the Mafia boss involved was a neighbour of the owner of the Blackstock Lumber Company.

The HTA fires stopped around 1991. No one knows whether this was because the Mafia boss, if he existed, had disposed of an associate who had become too troublesome or had simply stopped this particular racket. Unfortunately for the sleuths, soon after the HTA fires stopped, the city was hit by a far more widespread and dangerous outbreak of fires, over a hundred in all, costing more than $35 million in damage and accounting for three lives. The epidemic ensured that no resources were available for any other investigation, including the HTA fires. All of the recent fires had been set by a young arsonist, Paul Keller. In jail he admitted* that he didn't know why he had started them, and indeed sometimes forgot he *had* started them until he saw them on the TV news bulletins. Nevertheless, on some occasions, he stayed to watch the blaze: 'Mainly what I saw was how incredibly fast they took off. They burned so fast they blew me away.'

Keller, an advertising salesman and member of a church choir, was not an obvious arsonist, if only because he looked so respectable and dressed so well. His year-long spree started in August 1992 when he set light to three half-built houses, moving on three days later to two churches. In a fashion typical of serial arsonists, he soon grew bolder, setting off one blaze, then waiting till the fire engines had gone out before setting off another near the premises they had just left.

Within a few months, however, the clues started to mount. The fires were never set in bad weather, because Keller did not want to provide clues by going around in dirty clothes, witnesses had seen the sort of cars typical of salesmen near the blazes – and he had used his credit card to buy petrol near where the fires had erupted. But he was caught only when his father recognized his son from a composite picture put out by the police. Keller pleaded guilty to thirty-two counts of arson and admitted forty-five others. He was sentenced to seventy-five years in prison, which was increased to ninety-nine years after he was found guilty of setting fire to a retirement home and causing the deaths of three old women.

*In *USA Today*, 25 March 1994.

And why did he do it? Perhaps because he had twice been rejected in attempts to be a volunteer firefighter – or perhaps because, as he told a TV interviewer, he had been sexually abused by a volunteer firefighter when he was twelve years old.

# 8
# The Most Bullied Arsonist

As scientists we work in a world of probabilities and statistics and whilst the effort, the amount of work one puts into it, one always tries to be right, one always tries to provide for the court the most logical explanation and one always thinks that is the right explanation. I don't think I can always be right and it would be very nice if sometimes somebody who's actually set fire to something that I say is arson actually comes up after the court trial and says yes, you were right, that would give me that extra bit of confidence.

> Bob Bell, senior forensic scientist, Forensic Science Service

In February 1996 Blue Watch at the Speedwell Fire Station in Bristol was called to an alarm at Leo's Supermarket at Staple Hill. Leo's was not an ordinary supermarket. Originally it had been a terrace of houses, which had been converted into a small department store, mostly single-storey but with a two-storey addition. By the time of the fire it had been transformed into a supermarket: there was a furniture store on the corner of the ground floor, a tyre salesroom and the second floor had been turned into a club. It was a rambling, complex structure.

If Leo's was out of the ordinary, so was one of the firefighters, the twenty-year-old Fleur Lombard. She was one of the most recent recruits to the station and her close friend Sue Osbourne remembers that 'She was very confident. I couldn't believe someone so young could have that confidence. She was going to break the mould, she was going to go up through the ranks. She wanted to be the first female chief and she would have got it, if she hadn't died.'

Her opinion is shared by Peter Shilton, a senior firefighter who had been on the panel that had selected her. 'Fleur was an extremely bright student during her recruit training – you don't get the silver axe award [for the best student in the group] without a great deal of effort and certainly she showed that in abundance. It became clear, as she progressed through her training and beyond, that she had a tremendous future in front of her and, undoubtedly, would have risen to the highest levels in the fire service.'

Fleur was the quintessential post-feminist. As Osbourne tells it, 'At the end of our training school because we were the first four women in Avon to be recruited we were on the front page of our local newspaper. She was very angry about that. She said there were eighteen of us in that training school and why were only four on the front of that newspaper? She was angry that we were given the publicity while the blokes worked just as hard as we did.'

At first the firefighters thought that the fire at Leo's was a routine blaze. But then they learnt that people might be left in the supermarket. 'That changed the way the fire was fought,' says Rob Seaman, Fleur's colleague. 'If there's nobody in a building and it's quite a serious fire then you tend to fight it defensively rather than offensively because you're not putting the crews at too much risk. But if there's people inside, or you think there's people inside, then it has to be fought offensively, so you send guys in to tackle the fire and to search as well, and that's what happened. Fleur and I were to lay the guideline into the supermarket and our other team were to take a hose-line in.'

Unfortunately, says Peter Shilton, 'because of some confusion over the briefing that they'd been given the crews separated almost immediately they entered the building. One of the crews went straight ahead and the other crew, wearing breathing apparatus, who had been ordered to lay a guideline, proceeded towards the left-hand wall of the building.' That team was Fleur Lombard and Rob Seaman. 'They continued down along an aisle of the store laying the guideline as required. In the meantime the crew that had gone straight ahead had decided, after three minutes inside the building, that they needed to retreat. They began to walk back towards the main entrance from which they'd entered the building.'

Seaman and Fleur Lombard had been searching for survivors. 'By the time we got down to the end of the aisle into the corner,' says Seaman, 'the noises really started to happen. Lots of bangs, lots of explosions. The smoke had come right down and the heat started to build up very rapidly. I was trying to tie off in the corner of the store but I couldn't really find anything suitable to tie off to – they've got to be substantial tie-off points so that the guideline doesn't move. We were both aware of the amount of heat that was building up and the noises. It was getting louder and louder, which was the most frightening thing because we weren't really sure how close it was.'

Suddenly a series of explosions tore through the roof of the store, caused by a flash-over of the fire in which temperatures had soared to 1000°C. 'By this time,' Seaman remembers, 'we were down on our knees and I shouted to Fleur that we couldn't stay in here much longer. A few seconds later Fleur shouted to me, "Evacuate, evacuate!" so we turned, I grabbed hold of her waist belt, she grabbed hold of mine, I can't remember if I finished tying off the line – I mean that sort of paled into insignificance at the time. So we proceeded to make our way out, keeping very low. The noises were getting closer and a lot louder now, the smoke level was completely down, you couldn't see anything, and there was a glow rolling above us and down the sides and a little bit along the floor. We just wanted to get out of there. I don't know how far we had travelled but then something happened that I still can't explain today. Then I sort of came to and I was face down on the floor for a few seconds wondering where I was. Then I could feel the burning and the heat and the noises again and then suddenly realized where I was and what I was doing.

'I don't know what happened then – adrenaline took over and I ran around for a few seconds. I scrambled around a little bit more in the right direction and then I was grabbed by a firefighter. I asked if Fleur was out and they said no, she wasn't. I ran around for a couple of seconds wondering what to do, and then I just grabbed a hose reel off one of the guys and I went back in to try to find her. I saw her straight away and we just got to her as quick as we could. I picked her up underneath her arms. Her helmet was gone and her BA just dropped off her where the straps had burned through. Pat grabbed Fleur's legs, I grabbed her body and we got her out as quick as we could. I remember putting her down, and after that I can't remember too much until I was in the ambulance.'

Fleur's death was traumatic for the whole brigade. 'Everybody was in total disbelief that one of our own firefighters could have lost their life in what was a relatively straightforward job,' says Peter Shilton. 'The shock waves were felt throughout the British fire service. Everybody seemed to share this sense of shock that we in Avon were feeling at the time, total disbelief that it could have happened in these circumstances.'

Inevitably it was worse for Sue Osbourne. 'The day Fleur died,' she remembers, 'there was a stunned silence. It was almost as if the brigade had gone into shock. She was so young and she was female. People couldn't believe it, it was just stunned silence. You'd go into fire stations and people were just quiet, it was horrendous, it was awful.'

Weirdly, Fleur seems to have had some form of premonition of her death: 'Four days before she died,' Osbourne remembers, 'we were just pulling into the Asda car park and it came on the radio about two firemen from Blaenau that had died. We listened to the report and then she turned round to me and said, "Oh, you don't expect to die in a house fire, a supermarket maybe", and she nodded to Asda superstore.'

Fleur's funeral was the worst day of Sue Osbourne's life. The only consolation was that the public shared her grief. 'The day of the funeral it was freezing cold,' she recalls. 'Thousands of people all stood along the road while we walked down with her coffin on the turntable ladder. We'd choreographed every footstep to get her coffin on to our shoulders and it was just horrendous. There were people crying, and it was just bitter cold. I was just stood there with my best friend on my shoulder and she was dead. I wanted to cry and I couldn't. I had to keep in my head what footstep I had to make here and what footstep I had to make there and I didn't want to let her family down.'

The posthumous award of a George Medal, the country's highest award for civilian bravery, to someone who had been a fully fledged firefighter for a mere three months was almost automatic.

The state of Fleur's clothing and equipment gave Shilton and his colleagues some indication of the ferocity of the blaze. 'Fleur's breathing apparatus was discovered the following day. Most of the combustible components had been destroyed and her helmet was never found – it was totally consumed by the fire. This suggested that the temperatures must have been exceedingly high. Fleur died as a result of a massive flash-over and the temperatures estimated were in excess of 1000°C, probably more in the order of about 1200°C. That exceeded by at least four times the design specification of the personal protective equipment and the breathing apparatus components, and they had been badly affected in those temperatures. She died instantaneously, she wouldn't have felt anything, and it was a phenomenon that could not have been predicted or foreseen.'

Because Fleur Lombard was the first female firefighter to die on active duty in Britain the investigation into the cause of the fire became an emotional business for everyone concerned. 'This was the death of one of our own,' says Peter Shilton, 'and it took on a whole new mantle. We were determined to leave no stone unturned to find out exactly what had happened, so that we could learn lessons from that and hopefully put things right for the future.

'One of the things that we were very conscious of was the traumatic effects on firefighters. Because so many interested agencies wanted to undertake their own inquiries we agreed between the Health and Safety Executive, the Fire Brigades Union, the Home Office and the police that we would minimize the individual questioning of firefighters, combine them all together so that the inquiries would be co-ordinated by us. That worked well and minimized the stress and trauma that the firefighters had to endure.'

The next day when Peter Shilton visited the scene he 'saw complete devastation. The store itself was gutted and there was virtually no part of the roof structure still intact. The roof had collapsed into the building, all of the partition walls had caved in. All that remained was the outer shell, which contained huge piles of black debris, the remains of the stock from the store.'

Bob Bell, of the Forensic Science Service – with his grey beard and long grey hair the very model of the brilliant, if slightly eccentric scientist – was the first investigator to sort through the debris and try to reconstruct what had happened. 'We'd had a sprinkling of snow,' he says, 'and this, of course, meant that when we walked in, instead of everything being black, as one normally expects at a fire scene, everything was white. I suppose that immediately caused me to look at everything in a different way. Unusually the first thing we had to do was to try to warm the place up, get rid of some of the snow without disturbing what was underneath.

'It had been a large shop, in terms of floor area, and had been constructed out of a number of separate buildings, which had been joined together and the fire had spread through all of them. The shop was absolutely devastated: where the counters had been were merely twisted masses of shelves.

'The most fire-damaged area was clearly in the centre of the building but I had information that the fire did not start there – there were shoppers in the building when the fire started and from witness statements and the police I knew that the fire had started in the meat preparation and storage area, which was to one end of the building. Had there not been people in the shop, the fire investigation would have been very much more difficult because knowing which way the fire had spread was crucial to being able to go back and pinpoint the seat of the fire. Had it happened in the middle of the night and been burning for an hour before anybody knew anything about it then we might have started the investigation in the centre of the shop and gradually had to work outwards. We could have been there for weeks sifting through it all until we came to the seat.

'The meat preparation area was cut off from the rest of the shop by some stud-work.' On the other side was the equipment. There was also, says Bell, 'an area that had been divided off. The shop had previously had a deli-catessen counter, with a decorative hood. That was where the smoke was first seen. They put chocolate bars and that sort of thing over the counter, to cover it over because they were no longer using it and behind it they'd stored packets of crisps.'

Bell immediately spotted this was the logical place for the fire to start, 'because obviously we had a quick fire. You needed something to build up some heat, to get going quickly, to move actually out into the shop.' But once he looked carefully he found 'that the fire had spread into there from the rest of the meat preparation area. It hadn't actually started there.' He soon realized that 'the cold room had collapsed because the fire had attacked it from one corner' – and because there had been some heavy machinery on top. 'Now that suggested straight away that if it had spread into there then the fire was going well by the time there was any indication in the shop that there was a fire.'

Bell ruminated about possible causes. 'Looking around you could see cables hanging down so there's a possibility of an electrical fault. There's obviously the remains of packaging material so there's a possibility of a dropped cigarette or arson. There was a large chopping block right in the centre where a lot of the meat preparation took place and this had been burned. You could see the shadow side and the side where most of the fire damage had taken place and that immediately points you back in one direc-tion to where the fire had probably begun. Look along the wall, and clearly the fire had spread from left to right along a workbench. At one end of the workbench there's low-level heat, the workbench had collapsed. Why did it collapse? What was happening at that end that wasn't happening at the other end? A clear explanation is that there was more heat, more fire, lower level. Once you get to somewhere where you've got low-level fire you're probably homing in on the seat. Nevertheless the other workbench at right angles to it had collapsed, disappeared completely. There were the remains of electrical equipment so I had to look at these, take some of them back to the laboratory, look at some on site. Some are straightforward, some have clearly got the cable over them – they weren't even plugged in. A piece of electrical equipment that isn't plugged in is unlikely to cause a fire. Right in the corner there are cables coming down the wall, right at the bottom there is what looks to me like a junction box, which is not particularly well made. The connections aren't quite as large as I might have expected but, you know, that in itself won't cause the fire.'

Another lead that led nowhere was 'a cable duct that came down the wall. This was right at the end of the bench that had collapsed and I know for a fact there must have been low-level burning there. I took a very detailed look at those cables, those connectors. Had there been arcing, had there been long-term wasting away of the conductors, a sure sign that something was wrong? As much as I looked at that, no signs of a fault, none at all. . . . True, some of them had been on at the time of the fire and when the fire burns through the insulation, yes, you get some arcing but nothing that could indicate a fault in that area. Moving along to the other side, there were the remains of some sockets, which had been in the wall that had collapsed. I looked at those, no indication of a fault, looked at the machinery, some of which was quite heavy. There was a sausage-making machine, made out of really heavy cast-iron, a really solid job, not something that can readily over-heat and cause a fire, but there were electronic weighing machines, electronic pricing machines. Were these the cause of the trouble? I look at these, I look at the contacts in them, look at the wiring, any indication at all and, in fact, no, there wasn't.

'In my mind I was coming down to something low-level in the corner of that room but not the wiring itself. Now, in the corner there was a partition that divided the meat preparation area from the shop itself and of course the partition was no longer there. On the other side of the partition were some deep freezes. Now it is known that they had had some problem with some deep freezes in the shop before.'

As he searched Bell realized that the source of the fire 'had to have been in the meat preparation area itself. Clearing all the debris away, came down to low level, the remains of cardboard boxes, the remains of pricing rolls, the remains of packaging. Now, this is supposed to be a reasonably clean area but everybody has their little pile of bits and pieces so it was important at this stage to get in the manager of the butchery area who would know what was in it. In talking to him he quite freely said, "Oh, yes, there were several cardboard boxes there. We had polystyrene trays there for holding the meat products." That was the clue I required, that was the piece of information that told me not only could the fire have started there from everything that I could see but also the identity of the fuel that created the heat necessary for the fire spread. So in my mind at this stage I have a seat and I have some fuel. The only crucial thing I do not have is a source of ignition, what caused the fire.'

He also needed to know 'whether in fact this was an accidental fire or whether it had been deliberately set. I needed to know whether somebody in that area had been smoking – it's a non-smoking area but had somebody been smoking in it? It's the sort of question that I need an honest answer to but people aren't that interested in giving me one. But the manager was certain that no one would have been smoking in that area. So I am now thinking we have got a serious fire which has begun in that corner, everything points to that, so what could have caused it? I come back again to the electrical wiring. Am I right about that? Another look. Yes, I'm convinced it was not that. Was it anything on the bench by the side? No, it started here in these boxes of polystyrene trays. I am becoming convinced about that because everything fits. What could have ignited it? It is very unlikely that anything but a naked flame would have set that on fire under those conditions, too far away from the cabling and I can't see anything that would ignite it from that. The machinery isn't at fault. It didn't come through from the deep freeze. We're left with only one possible cause. Naked flame. That means straight away that somebody did this deliberately.'

Once Bell had decided that the fire was the result of arson the police swung into action, concentrating on those who had been near the seat of the fire at the time. No accelerants had been used (if they had, a suspect's clothing would have shown signs of it), and there hadn't been a break-in.

The debris was examined and the packaging was tested by Penny Morgan, senior fire consultant, Centre for Fire Protection. She'd gone to see if the fire meant that there should be changes in the building regulations and to find an answer to the reason why only one of a two-person crew had survived the blaze. 'Knowing how closely they work together,' she says, 'that in itself is unusual. We wanted to see if anything unusual had occurred within the building.'

When she examined the contents of the meat preparation area in detail she found that there were 'polystyrene trays stacked up in great piles on one of the tables. There were also boxes of meat pads – the little pads that go under the meat in the trays to soak up blood and water. When you buy meat, you have a tray, a pad, the meat and then clingfilm wrapped around it. Now, all of these are combustible, and because some of them were left out ready for work on Monday, it was possible that this might have been the place where it started. Because we've got pale smoke we are then suspicious that either the trays or the pads may have been the material first ignited.' She took small similar samples from another branch of Leo's and tested them in a 'cone calorimeter', which measures the release of heat. Larger samples were used to see if the pads gave off the pale-coloured smoke noted by those on the scene.

Pressed by the police – because normally her role did not include any involvement in forensic work – she proved that the fire could well have grown quickly in a box of meat pads: 'Sufficient that, yes, you can say here's the start of something that's going to spread.' Then she set about charting the spread of the fire. She concluded that 'Once the fire had started in the meat pads, we established that we've got enough energy there to get a big enough fire that's going to involve other materials present. We have other polymers in the meat prep room, boxes of crisps nearby, and we have all the stuff on display in the shop. When the smoke is produced, it is made up of unburnt material and hot gases. It's buoyant, it's moving upwards and away from the source because it's short of oxygen. The smoke from that original fire in the meat preparation room would have moved up, back into the storage area and forward into the retail area.

'Once the smoke and hot gases have started moving, they will start to head up the ceiling under which they're moving.' And it was here that the higgledy-piggledy nature of the building provided a major clue to the nature of the fire, the speed with which it spread and, indeed, the mystery of the death of Fleur Lombard. Penny Morgan was particularly interested in the roofs of the extensions, 'what materials had been used in them, because what was obvious to people working in the shop, and certainly what would not have been obvious to the brigade, was what sort of lining materials were present. We found subsequently that the false ceiling we would normally have expected to be of plasterboard – a non-combustible material – was partly fibreboard. The ceiling had been made to look as though it was tiled with wooden strips but it was large sheets of fibreboard.

'When fibreboard is warmed up it releases volatiles from the glues, adhesive and all the other gubbins holding it together. Hot gases are moving about, heat is radiating down, igniting material elsewhere, which in turn is producing more smoke. You have something that's generating, pulsing, turbulent, and very fast-acting. So the time taken between the witnesses seeing the wispy smoke and a member of staff checking to find that they could barely get out of the building because the smoke was so deep was probably less than a minute. Now, we know we've had a very fast-developing fire, probably because of the amount of polymer present in the initial area. Polymers are notorious for burning quickly, giving you high heat release fast, probably quite short duration but it's enough energy, hot gases, smoke moving away from that area to involve other materials. The thing becomes self-propagating until you can fight the fire.'

As Morgan points out, the arsonist had chosen the ideal spot to start a major conflagration. 'The meat preparation room was out of sight of everybody. It contained a lot of combustible material in the form and layout, there was insufficient space for the fire to develop undetected. It could spread out of it easily and involve other materials.' The arsonist was also lucky because nearby 'was a sort of temporary store filled with boxes of crisps. They burn exceptionally well – they're carbohydrate and fat, and they're in an airtight container, which is usually highly flammable. It tends to be waxed paper or some sort of polymer – the foil versions are the best if you want to burn them, very easy to ignite with a match or hot gases, they will go at quite low temperatures. Once they do, an individual bag will sustain burning for say half a minute or so, and will transfer the heat quite rapidly and give you quite a big fire very quickly. It'll give off quite a lot of dark smoke and quite a lot of flaming, very high temperature. I would estimate you can get something like the equivalent of an armchair from perhaps two or three hundred packets. It was a perfect spot. If you wanted to start a fire that was the place to go.'

But, says Morgan, 'It's unlikely he would have realized just how serious that fire would be. It's unlikely that most arsonists – unless they really want to destroy a building – have any concept of what it is they're actually starting. Having chosen to drop a match, we suspect, into the meat pads, the fire would have maybe just started, he'd have seen it, could have closed the door on it and walked away and would have expected there to have been a small amount of smoke. Witnesses saw a small amount of smoke. He was sent to investigate. The trouble was, by the time he got there it was already getting too big and it wasn't something you could tackle. This is a lesson for all arsonists, that they may in fact kill themselves and often do but they may also kill other people.'

There remained the problem of why the flash-over had killed Fleur Lombard but not Rob Seaman. 'They were both at risk,' says Penny Morgan, 'but only one of them died. Now that indicated to me that maybe there was something other than a straightforward flash-over. How come we'd not had a double tragedy?' Theoretically, in a flash-over all firefighters are equal. 'Flash-over is a critical time when anybody present in the building can be put at risk, regardless of the fact that the brigade wear heavy protective gear. The BA crew have their own atmosphere to breathe so they're not going to breathe any hot gases. But if there's a flash-over then anybody in the building is equally at risk. In this case we have a BA crew who worked very closely together and they moved through the building more or less in contact if not *actually* in contact. The risk for them was equal but Fleur was caught in something and died and Rob was caught in something but survived. Now that seemed to indicate to me and my colleagues here at Fire Research that something else had happened, and we wanted to get a feel for the scene and see whether this was just a version of an ordinary flash-over or something else.

'Now, remember, it's very hot, dark, smoky. There's no way they could tell that something was happening at ceiling level above them, that's going to be nearly three metres above ground level. Once flash-over was inevitable in the building, because of the involvement of all the material on display, then the geometry of the building became important. In this building we had an L-shape and we also had difference in the height of the ceilings. It was possible that you wouldn't have a straightforward flash-over that just filled the whole area, it might have been slightly earlier than that where we had those hot gases moving at high speed but were then affected and directed by the way the building was laid out, by the fact that part of the ceiling level was slightly lower, so that you had movement of the hot gases not straight and back but straight and down. So you could have just a very localized effect, and this is the one we examined. We looked at the plans and found that if you placed a person in a particular spot at the far end of the building, quite close to the entrance but the far end from where the fire started, and if you looked carefully at what was present you found you have just the sort of geometry that would allow for a local effect and that, we think, is what happened to Fleur. She was caught in a local deepening effect we call it, where those hot gases are locally directed. She was caught in hot gases that were probably at flame temperature, probably in excess of 800°C.' She was only caught for a short period of time, maybe only a few seconds: 'A flash-over is very fast but a few seconds in hot gases at flame temperature – and she was very unlikely to survive. That would explain why she died and Rob survived, even though he was very close by.'

Morgan came up with the tragic conclusion that 'Fleur was in the wrong place at the wrong time.' Even worse, although the fire was unpredictable for the firefighters present, 'it should have been avoidable because somebody should have realized that the fibreboard should have been replaced. There had been an edict from the department about the use of fibreboard in 1968 – though not in commercial premises. It was well known.'

While Penny Morgan and her colleagues were teasing out the details of Fleur's death, the police were interviewing witnesses – including the store's security guard, twenty-one-year-old Martin Cody, who had started work at the supermarket that night. Witnesses described how they had seen Cody punching the air, while watching the fire, as if in celebration, and boasting that he knew that the fire had started in the meat preparation area. Later during the fire he had been captured on video, nonchalant, unconcerned, although he had known by then of the death of a firefighter.

As the investigation concentrated on him he was revealed as a disturbed personality. He had first attracted media attention six years earlier when his mother had claimed that he was the most bullied schoolboy in Britain. When he left school two years later fire had followed him from job to job wherever he moved. In June 1997 he was convicted of the arson and of the manslaughter of Fleur Lombard.

But Fleur Lombard did not die in vain. As Peter Shilton points out: 'One of the positive things that came out of the tragedy at Leo's was recognition of our inadequate training facilities. As a direct result of that we've now been able to provide a brand new hot-fire training centre, which will enable us in the future to train all our firefighters about dangerous conditions where heat, smoke and the threat of flash-over are present. This is an exciting new era for us and will take our training into the next century.'

# 9
# The Philadelphia Story

Every time you determine a fire to be an arson you always have great expectations to solve the case and determine who set the fire, it doesn't always hold true but you always start out with those expectations.

Lieutenant Tommy Lawson, Philadelphia fire marshal's department

It was typical of many American cities that over half the fires investigated in Philadelphia in the early 1990s – fifty a week – turned out to have been set deliberately. Fewer than one in five of these arson attacks resulted in a conviction but one of the most significant exceptions was the incident known as the 'I-95 fire'. Interstate 95 is the main north-south highway down the east coast of the USA. It is also the main artery through the heart of the city of Philadelphia. On the morning of 14 March 1996 the citizens awoke to find that a fire had shut down I-95 until further notice. As a result the city went into gridlock.

Within minutes of their arrival on the scene the previous evening, the crew of Engine 28 of the Philadelphia Fire Department concluded that they couldn't handle the blaze alone; in the end 150 firefighters were involved in trying to put it out. As Lieutenant Tommy Lawson walked east towards the fire, 'I could see that a large volume of fire was located underneath I-95. I could see a great deal of structural damage had occurred to the highway. Concrete was broken loose, steel support rods were hanging.'

The fuel for the fire was Philadelphia's biggest stack of used car tyres – almost a million of them – dumped alongside an elevated section of the highway. 'It was at least two storeys in height,' says Lawson. There was a large hole in the fence surrounding the pile and Lawson 'was able to make my way through that hole into the area under 95. What I found was the smouldering remains of a large volume of tyres. Most of the rubber had

111

been burned away. The only thing left was the steel belts, which at that time was just a large pile of tangled wire.' This was not surprising. 'Tyres are not easily ignited: they need a little boost to get them burning but once ignited they burn tremendously, they give off a great deal of heat and the BTU output* of a burning tyre is similar to gasoline so each one was almost like a gallon of gasoline burning. With the burning rubber and the tyres with their petroleum base, you get a real thick black smoke. It almost has a gritty appearance to it and, of course that was all over the neighbourhood. Everything you touched was black, everybody that was at the fire scene that night was covered in soot and smoke, the apparatus was black, it was one of the dirtier scenes that I've been to.

'There were several times when I thought the fire was going to be contained and each time I was wrong. It just grew in intensity and involved all of the tyres at this area. It eventually consumed a three-storey building adjacent to the tyres, two one-storey buildings were also involved and we were there most of the night. We had plenty of water but we just could not reach the seat of the fire. Consequently the fire kept extending to more and more tyres. There were several times during the night when it appeared as though we had the fire well in hand, the smoke was turning white and I felt that we had got it now. . . . But every time the fire just got bigger again and finally involved all the buildings that were there.' Whatever they did, the fire 'continued to burn underneath. For several days afterwards we had companies there pouring thousands of gallons of water on it and the smoke never lifted for a couple of days. It was a pretty bad scene.'

Nevertheless, as Steve Avato of the ATF points out, 'We were lucky that the wind was blowing towards the Delaware river. Had it blown the other way houses would have been involved and a lot more destruction than actually occurred.'

From the start, says Avato, they assumed it was a case of arson. 'Typically tyres don't ignite on their own. My experience with tyre fires prior to that was that they were almost always arson.' Lawson agrees: 'The absence of an obvious ignition source, the fact that it was an isolated area, not an area where you would find someone walking at night. It's very dark, desolate, the area is used to dump debris and the difficulty of igniting automobile tyres – all these factors combined, at least in my mind, to say that we

*British Thermal Units – a standard measurement of heat.

had an arson.'

The desolation of the area meant that conditions were far from ideal for investigation. 'It was already late at night,' says Avato, 'it was a deserted area of the city so our chances of having witnesses were pretty slim.' Lawson concurs: 'It involved an area where there were no houses, no foot traffic, no vehicular traffic. If you have a fire in a house you usually have a story from the occupants of the house that gives you a starting point. This appeared to be a random fire for no rhyme or reason.' What made matters worse was the pressure on the investigators because of the size of the fire and because it was so visible from I-95. A task force was formed, from the Philadelphia Police Department, the Fire Department and local agents of the ATF.

In situations like this the investigators rely on thoroughness and persistence. 'In looking at a fire scene, particularly a large scene like the I-95 fire' says Lieutenant Renney Pelszynski, 'we don't always know what we're looking for exactly. You're taking in what the scene encompasses, looking for things that don't seem to fit – you never know exactly what that is until you see it and that's typically based on your experience as a firefighter or a fire investigator. In this case we were looking for things that just didn't seem to fit into a tyre-storage facility. For example, there were vehicles mixed in among the tyres, there were pieces of equipment, there were cutting torches. We didn't know what belonged, what did not belong. We found fuel containers, portable gasoline cans – did they belong in the scene, were they part of the typical operation of this particular business or had the arsonist brought them into the scene and left them? Those are the types of things we were looking for.'

Their only advantage was that they could see where the fire had started. Pelszynski saw the 'underside of the I-95 roadway. The actual amount of damage to the underside surface was phenomenal: there was cracking of the concrete, sections had broken away, it was obviously seriously damaged and it appeared that it would be shut down for quite some time pending its repairs. This was confirmed by interviews with the firefighters.' Moreover, from the start it seemed probable that the arsonist had got in through the same hole in the fence that Lawson had climbed through.

Immediately Lawson started looking for an ignition source for the tyres but 'the only thing I found was the remnants of tyres. The twisted steel belts were still there, some evidence of rubber, but most of it had been reduced to powder and I canvassed the area for any large accumulations of rubbish.

While there was a lot of rubbish in the area it was scattered about and there were no large accumulations that would provide a good ignition source for the tyres.

'Of course, there was no accidental ignition source located under there. I started looking for any type of containers that may have held an inflammable liquid – the street was covered with a lot of containers. There was a lot of trash on the street but not piles of trash that would be an ignition source for the tyres. I went up and down the street checking all the containers for any odour of an accelerant of any kind, particularly gasoline. It was my feeling at the time that the fire was initiated with an accelerant.' It was not agreeable work. The containers were not, says Lawson, 'the worst thing I've ever sniffed. When you're examining a fire scene everything you pick up is not always nice to pick up. I did find some old oil containers with what appeared to be motor oil but I found nothing to indicate that a container held gasoline.'

The next step for the investigator 'is to start conducting interviews to find out who may have seen or heard anything. A fire of this size certainly draws a crowd and you start working the crowd, start going through, handing out business cards. Sometimes people don't like to talk to you in front of other people but if they have your card they'll give you a call later on when they're by themselves. We tried to touch base with people in the crowd, photographed the scene, documented the progression of the fire and just tried to develop any leads that we could.'

The most obvious suspect was the man who owned the tyre dump: Danny Carr, plump, grey-haired, imperturbable in his immaculate grey suit, white shirt and thick-rimmed glasses. He had been told by the Police Environmental Unit to clean up the site and remove the tyres. A court order had been served instructing him to do the necessary work before 14 March, the morning after the fire. That, says Avato, 'may give someone a very good reason for wanting these tyres to go away and one way to get them to go away would be to burn them'. Typically, Avato's language was cautious because, he says, 'As this case clearly illustrates, there's other possibilities and before you make that conclusion you have to be absolutely sure that all the evidence points to that particular individual. As an investigator you really don't want to jump at the first suspect you're given: even though the information looks very good there are other possibilities, and you have to be sure before you arrest someone for setting a fire.'

But even the ultra-cautions Avato has to admit that it was 'sort of an exciting thing, to show up at a scene and have someone almost handed to you as a suspect. It was almost a gift but you can't just take that and say it absolutely had to have been him because you need other facts to corroborate that. We worked continuously from the time I arrived on the scene for thirty-six hours straight without a break. Some time just before midnight of the same day Mr Carr was arrested by the Philadelphia district attorney's office on charges of causing or risking a catastrophe and some other state charges.'

To establish their case, says Avato, 'We made contact with members of the Philadelphia Police Department who had been involved in the investigation of Daniel Carr, prior to the fire, when he was cited for the illegal tyre dump. We obtained from them a list of Carr's employees.'

'In two particular employee interviews that we conducted shortly after the fire,' says Renney Pelszynski, 'the employees themselves indicated to us that Mr Carr was setting the building up to burn. They had moved at his direction a large quantity of gasoline into the property and they told us specifically that they believed that Daniel Carr was the responsible party for the fire.'

But all this was pretty circumstantial stuff. 'As we began to dig through what was left of the tyres,' says Avato, 'we found some evidence that corroborated what the employees had told us. We found a fuel tank that had been covered or surrounded by tyres, but there were some other things we found at the scene that didn't seem to make sense. For example we located a large tyre-shredding machine and when we checked on this machine it turned out that Mr Carr had recently purchased it and that the purchase was financed. We checked to find out if this machine was insured and it turned out the machine was not insured so it seemed rather odd that the owner of this business would burn his own business and allow this very expensive machine to burn up in the fire when he had no insurance to cover that loss.' They also tracked down 'an electrician who had done some pretty extensive electrical work in the building just days before the fire. If you were planning on burning your business or all of your tyres why would you spend the time and money to have an electrician do this work? These things started to bother us a little bit and didn't make sense.'

When the investigators put the facts together, 'It came to a point,' says Avato, 'where the consensus was that Daniel Carr did not set this fire.'

Inevitably, 'there was an emotional letdown. That was our best suspect and now he's gone and so you have to start at square one again. We had to look at the entire area around there, what other fires had been in the area, go back and re-interview witnesses, try to find other suspects and other reasons for the fire. That's exactly what we did in this case. It was almost like an emotional roller-coaster. We started out very high with a very good suspect initially, the information came in and then it looked like he had nothing to do with the fire and all of a sudden we have no suspects and it was time to try to find others.

'The thought crossed our minds that we may never have other suspects in this case. We had spent nearly a month of investigative time, ten to twelve hours a day, trying to pursue leads and all we got out of it was that one person didn't do it, and that left a whole group of other people who could have done it, unknown to us, and maybe we wouldn't figure out who it had been '

Because the I-95 fire had made such an impact, investigators like Pelszynski 'wanted to solve this, probably a little more so than most fires and there was a sense of frustration when we had eliminated our prime suspect. We knew we had to go in a new direction, we had to pursue another motive and more suspects, and we did. The community in which the I-95 fire occurred had been known to us for some time as a very busy area for fires, for arson fires particularly. Unfortunately a lot were set in commercial areas. There were few people in the area during the hours of darkness so we had little witness information to go on.'

As Steve Avato puts it, 'In a fire like that it could be the random act of an individual who just happened to be wandering by and saw a target of opportunity and set the fire. It could have been part of a pattern of other fires in the area. We re-interviewed firefighters, and there were firefighters who seemed to believe that there was a rash of fires in vacant buildings, that they had been to other tyre fires recently, so we began to look in that area. We started pulling the fire department's response logs to see what types of fires they were responding to, not knowing at that point that there would be any correlation to anything. We were looking for clues anywhere, we were looking for that needle in the haystack, so to speak.'

'We observed that one particular engine company, Engine 28, had been experiencing a tremendous amount of fires,' says Pelszynski, 'much more so than they typically had, and these fires were occurring generally over week nights. They were all after the hours of darkness. We were seeing a pattern emerge here and we knew that it was very out of character for this particular area and this particular engine company.'

These fires, Avato says, in a section of the city graphically called Badlands, were largely 'nuisance fires, trash cans, the trash dumpster fires, vacant buildings in the area where they were relatively small fires. We also found some larger fires in vacant buildings and they appeared to be occurring in the same time-frame, at night, after 11 or 11.30 but usually before 2 a.m. or so. We were starting to see a pattern but at that point there was no one to put with that pattern. These fires may have just been a coincidence.' They felt also, says Pelszynski, 'that one or more individuals were connected to this series of fires and we began to focus strongly on looking for this group'.

The investigators' break came in early April with a fire in a disused factory in Coral Street in the heart of Badlands that nearly turned into tragedy: it threatened a number of homes in nearby Buckius Street, which had to be evacuated. Early on, Tommy Lawson looked through a window to see where the fire had started and he could then eliminate three-quarters of the factory as irrelevant to the investigations. 'It was still night-time, there was a lot of water in there, still some hot spots, but I was able to get inside and examine the area where I initially saw the fire and there were a lot of things that I noticed almost immediately. First of all the property had no utilities, there was no electricity, there was no gas, no water, there was nothing in there that would cause an accidental fire. So immediately I knew that the fire had to have human involvement and, of course, my next thought was, could we have an accidental fire caused by someone doing something illegal in the building? I had some initial indications from neighbours that some people had gotten into the building in the past and stolen the wire, stolen some of the plumbing from the building, so these were the things that I was looking for immediately.

'Most of what I saw that night indicated that anything of real value had already been taken from the building. There was really not much left to steal in there. There was a great deal of combustibles consisting of paper, old baled invoices and I found remnants of wood panelling and wood studding and, in an initial examination of the scene and following the various thermal burn patterns, it was my indication at that time that I was looking at multiple points of origin'.

His confidence was reinforced by later study of the 'burn patterns'. 'As fire burns it leaves a trail behind it,' Lawson explains, 'of soot patterns, heat patterns, char. These are the after-effects of fire, you can visually observe these burn patterns and follow them. Fire burns in almost a preconceived way – it takes the path of least resistance.' He started looking for the patterns outside the building. He could see that 'Discolouration is more evident over some windows than others. It indicates to me without even entering the building where some of the heaviest fire was located and it gives me a starting point once I enter the building. I also used a snorkel to get me up above the building so that I could examine the roof from above, which I did, and I could see the greatest amount of damage that was done to the roof area, which also indicated to me where some of the heaviest concentration of fire was located.' By following the fire patterns inside the building he was able to reconstruct the way that the fire had followed three separate paths traceable back to three separate points of origin. Moreover, at each point there were combustible materials. Paper, wood, small pieces of furniture.

When he 'interviewed an occupant of the end house directly behind the factory he indicated to me that, some time earlier, he had heard what sounded to him like a group of juveniles or a group of young males trying to force entry into this building. It sounded to him as though they were pounding on the windows or on the door and then it stopped. At some time later they returned and started again, and this time he heard a car drive away. Then he became aware of the fire.

'As we received more information on this fire,' says Lawson, 'it was consistent with information we had received on other fires that we were investigating in the same area. I got good feelings about this investigation. It's hard to describe . . . Sometimes you get an investigation going and almost from the beginning you feel like you're running into a brick wall. This particular fire, things started clicking and – it's hard to put into words but, it's just a feeling that you get where you say to yourself, "We're going to break this one", and it worked out real well. It's a whole series of events. First of all, this is not the first fire these companies fought that night. Just an hour earlier they were involved in a fire in a similar type of building that involved large piles of rubbish and lumber that were stored there.'

Lawson was also beginning to understand the type of people he was dealing with: 'If I had been responsible for that fire I would have been scared to death. I probably would not have been involved with another fire for the rest of my life. These people didn't stop, it was almost as if that was a big boost.

Shortly after that the Buckias fire destroyed people's homes, destroyed their lives, and the fires still did not stop. This told me that these people had no consideration for people's lives or property, they just didn't seem to care. That was real scary – anybody could have died in any one of these fires and it didn't stop them.'

Finally, Pelszynski says, 'We got a break from a police officer who observed a vehicle with more than one person in it. These particular fellows set a car on fire so now we knew we had to be looking for several juveniles or youths who may be responsible. Generally arsonists set fires alone – it's not typical that we see a group of people involved in fire-setting – so this was kind of abnormal for what we were used to. Initially I don't think we were looking at a group as much as perhaps one or two individuals working in concert and, frankly, I was surprised when it turned out to be a group of several individuals that were responsible for these fires. It wasn't typically what we find.'

They were also finding out more about the way the fires had been set: 'By an open flame application to the available combustibles. That basically means someone took a match or a lighter, set fire to those materials present. Then we observed that some of these fires were a little more sophisticated and some devices were used. There were rolled up *Auto Shoppers*, for example – a sales publication – at more than one of the fire scenes. They didn't belong at those scenes so we knew someone had to bring them into the property. That provided a common thread to these fires.'

Nevertheless it took a long time for them to accept that the enormous I-95 fire was within the pattern of fires they were investigating. In the end its very size was a clue: 'When we saw the pattern emerging with the car fires, the dumpsters and the vacant buildings, this seemed to be the type of fire where you have a thrill-seeker and the person is seeking a bigger and bigger thrill. They start out with dumpsters, they graduate to vacant buildings, cars, and what we were expecting was that soon this would involve occupied properties. Up to that point we hadn't sustained any injuries to anyone, and our concerns were that this would develop into something where someone would be injured or killed. We were hoping to stop it before that occurred.'

By then the investigation had identified a coherent pattern. What connected many fires like that in Coral Street, says Lawson, was the presence of witnesses who could 'at least tell us that they saw young white males leaving in a vehicle. Some of them were able to describe the vehicle as a white Oldsmobile.' But then a police officer had been able 'to give us a rough description and although we didn't have a tag number for the car involved we had an idea we were looking for a group of young white males in a white car that we knew would have to be a 1986 or later Oldsmobile' because several witnesses had noticed that it had three, not two, tail lights.

'A burglary detail from the east police division observed a white car matching the description of the car we're looking for leaving another fire scene and pursued it. They had marked police units make a vehicle stop, at which point they identified that five young white males were inside the car. There were backpacks in the car with them which contained two-way radios, a police scanner.' Moreover, says Avato, 'they were also monitoring the Fire Department channels and they were listening to the Fire Department activity as they drove around.' They also had on board what they later identified as butter bombs. (Pelszynski describes the devices as 'little egg-like containers that ladies' stockings were shipped in. They came in two halves and he [the arsonist] would fill them with polyurethane, then drill a hole and put a wick in. Crude devices but if you ignite the wick the plastic burns and that's what they used to set the I-95 fire as well as one of the vehicle fires.' It was a more sophisticated device than those the boys had used in other fires, which involved 'taking matches and lighters and setting fire to papers and available combustible materials' while 'professional arsonists generally will use inflammable liquids, gasoline, they'll use timing devices.')

Avato describes how these kids 'were stopped for further investigation and that is generally considered the big break in this case. We had an identification now on particular individuals who were at least in the area of the fire that night, they were not charged with the fire that night, there was no direct evidence there, but we knew now who we wanted to look into further.'

The next stage was to tie the fires to the five young men in the car, whose ages ranged from thirteen to nineteen. 'We found,' says Pelszynski, 'the connection between this group of individuals. They were actually members of a bowling team. We knew that they were together on the nights of these particular fires and it would have fitted a pattern where they had the opportunity to be out on the street at these fire scenes at the time the fires occurred. We just needed to put that together into kind of a neat package for prosecution.'

Lastly they had to get one of the suspects to talk. At that point Pelszynski and Avato had a break when one of the kids, a sixteen-year-old who turned out to have been the ringleader, started to get emotional: 'It became apparent by his body language that he was very nervous, he had something inside he wanted to talk about and didn't quite know how to go about doing that. We questioned him and watched his body language and it became apparent that he was going to tell us. So we did get him to confess to having set one dumpster fire. We knew that we had broken through, that we could pursue this. He's a young kid, he wasn't a hardened criminal', which made it much easier for them.

Avato continues: 'We had asked him about several small fires and he got very angry at us. He jumped out of the seat, began to curse at us and tell us that we had no right to come into his home and accuse him of such things. When he did that we knew that this was the reaction of a person who had been setting fires and he didn't want us to know that. We took that opportunity to bring the interview into our office where we could be in better control of it.' As Avato says, it was clear that 'he wanted to tell us but he was afraid of the consequences so it took a lot of reassuring to get him to that point. It is obviously a different interview strategy.'

'We needed to gain his trust,' says Pelszynski, 'we needed to make him comfortable in getting this off his chest and we pursued a line of questioning with him and he began to give up one fire after another. But it was clear that he still had more to tell us, there was something really troubling this kid, and at that point we didn't know that he was going to come out with a bombshell.'

So, says Avato, 'we started out by trying to calm him down and we began talking to him about some different fires in the area. Our strategy was to acknowledge that we knew he had set at least one fire but that there was another one that he wanted to tell us about. Finally he admitted that he and his friends had been involved in setting the I-95 tyre fire. Obviously inside we were excited about that but it wouldn't have looked right if we had shown how excited we were. We told him that we had known all along that he had done it and that he would feel much better now that he'd told us – and in fact he told us he did feel better. From there we got additional details on the fire itself.'

In the event the whole gang was convicted of setting thirteen fires, although the investigators believe that they had set at least thirty. It was left to Dian Williams, the criminal profiler, to pin down the rationale for the crime. 'Maybe this was a group of kids who were out and just raising hell.' So they were likely to be adolescents? 'They were male, because it's very unusual to find a female who's going to be involved in that kind of fire. They would probably find one ringleader, and the chances were very high that the ringleader would have absolutely no hands-on fire-setting but instead would direct the rest of the group to set fires. The ringleader would then be able to say, "I didn't do anything, I sat in the car, I tried to persuade them not to and nobody would listen to me."' As a group, too, their disregard for the consequences pointed to juveniles and, moreover, ones 'in a certain age group because there is so much immaturity'. They would also be 'kids who were not doing well in school and, in fact, kids who probably had dropped out of school or were truant a lot of the time and who were known as under-achievers. They would be kids who in general didn't have many friends. By the time they all came together as a group and decided that they were going to make a name for themselves by setting fires and getting away with doing this really dangerous and bad stuff, then that behaviour would reinforce for one another how smart they were. They were very bonded together, even though they didn't like one another, because they didn't feel particularly liked by anybody else, and they believed that what they were doing together in such a destructive way made them really special.'

Once they'd located the best place for a fire they would 'really get pumped up as though it was some major sports event, and go back then and under the cover of darkness set a fire, then watch it to make sure that it caught, then all congratulate one another and go and get a soda. This is very chilling, the deliberateness of that kind of fire-setting.'

In Williams' view they 'were not sorry that they set the fires, weren't sorry about the damage they caused and probably never will be. What they are sorry about is that they got caught.' And if they hadn't been caught? 'They would have continued to set bigger and more destructive fires because they were reinforcing for one another how much more clever they were than the Police Department, the Fire Department, any investigators who came along, ATF. They just laughed at all of that. Their motive was fame.' Williams is sure that they were headed for the criminal big league: fire-setting was just one of the ways to get there. 'Don't feel sorry for these kids,' she says. 'They were delinquent. They set out to do this really bad stuff, they didn't have any conscience about what they were doing, they're really asocial beings. They were having a great time, they loved what they were doing and they would have been delinquent if they weren't setting fires. They would have been doing something else but they chose fire-setting because it was exciting for them, it was a thrill, it was a game, you could be a star, the bigger the fire the greater the number of points. That was fun for them and for some of them the most fun they ever had, and isn't that awful?'

Only one person involved showed any sign of pity for any of the accused, and that was the unlikely figure of John McCool, who had lost his house and all his possessions in the Coral Street blaze. 'All through the trial,' he observed, 'one of the kids is sobbing his heart out and his parents were just so cold. I could see why he had no sense of hurting anyone because his parents had no sense of his hurt. It was killing me to watch this boy's heart break and his parents just more or less stayed back from him where I'd have been all over my kid. They'd have had to take me out in handcuffs before I'd let my son go. You know, you could understand where this boy was coming from, he wanted attention, he didn't care how he got it and he got it, unfortunately not the kind that I think he wanted. I just thought the boy needed a hug.'

# 10
# The Strange Case of the Pillow Pyro

There's numbers of investigators that he's worked with and trained and every one of them has been betrayed by what he did and it's never going to be put to rest to any of us until we know why he did it and he's the only one that can tell us why he did it and hopefully some day he'll find it in himself to tell us.

Mike Matassa, special agent, ATF

The case of the Pillow Pyro, as he came to be known, was almost certainly the biggest in the history of the large, fire-prone state of California, and took five years to solve. More to the point it was almost certainly the single most bizarre case in the history of fire investigation and one that, almost inevitably – not only for its inherent interest but also because it stopped so close to Hollywood – is due to be made into a 'major motion picture'.

The story starts in January 1987 when Marvin Casey, the fire investigator at the market town of Bakersfield, a hundred miles north-west of Los Angeles on Route 99, the inland road north to San Francisco, was called to a fire in a fabric shop. Strangely, he was immediately called to another fire, this time in an arts and craft shop. He was puzzled, not only by the coincidence ('we don't have a lot of fires in Bakersfield, you know') but by the fact that at the scene of the second fire, which fortunately had been put out almost immediately, he found the remains of a simple but effective incendiary device: a cigarette left burning with some matches in a rolled-up sheet of paper. Over the next few years it became clear that even such a small device could be relied on to cause immense damage.

Over the next few hours, though, Casey, a home-spun, honest, small-town detective, grew even more puzzled. There had been two fires in Fresno, during a conference on arson held in the town, and another in Tulare, another small town on Route 99 towards San Francisco half-way between Bakersfield and Fresno. Casey had the horrifying idea that perhaps the fires had been set by one of the fire investigators attending the conference. Fortunately, he had found a fingerprint on the device. It was sent to the local forensic fingerprint laboratory where it was treated with ninhydrin, a chemical that reacts with the amino acids found in sweat; the process turns any trace of a latent fingerprint purple. Unfortunately the result did not correspond with any held by the criminal records department and Casey had found that fifty-five of his colleagues – too many for him to investigate – had travelled home along Highway 99.

Two years later, when Casey had forgotten about the unsolved fires, there was another rash of blazes along Highway 101, the coast road between Los Angeles and San Francisco, also during an arson conference, this time at Salinas. Casey could narrow down the list of suspects. 'When I crossreferenced the two separate arson conferences,' he says, 'and came up with a list of ten guys who had attended both I was really excited.' However, the fingerprints of the ten did not correspond with the one found two years earlier.

Two more years went by. Then, between October 1990 and March 1991, Los Angeles County suffered from a spate of fires in well-known retail chain stores like Thrifty Drug Stores, Pier One Imports and Builders' Emporium. The fires were all quickly identified by investigators as deliberate and probably the work of the same arsonist, using the same *modus operandi*: parts of a home-made incendiary device were discovered at the scene of every fire. 'We had a series of fires in commercials,' says Glen Lucero, an investigator at the Los Angeles City Fire Department. 'That was unusual because the fires that were occurring in commercial properties were occurring predominantly during business hours. A lot of arson fires are set under the cover of darkness because arsonists are basically cowards at heart but to have a series of fires being set during business hours in businesses that were actually open was highly unusual.' It showed 'a certain amount of bravado and confidence by the person setting the fires'.

'These fires would occur in a spree,' says Mike Matassa of the ATF, 'maybe two or three fires within an hour on the same stretch of roadway. It was just too coincidental for them to not be related and we knew that we had a serial arsonist striking.'

What is more, says Lucero, 'they were following a certain pattern geographically. If you followed it on a map, for instance, you would see by putting little dots on the map that the person was following a specific area.' And not only were they close together, says Lucero, 'they were also occurring in close proximity time-wise. We had a series of fires that were occurring in a close geographical area and on top of that they were occurring within thirty minutes to an hour of each other.'

Fortunately, as Matassa tells the story, 'A critical break in the investigation occurred in late March at a fire that was the middle of five fires in a spree that day. It was at a small, medium-sized craft store and an incendiary device was recovered by one of the customers before it went off. It was the first time we found out what the device was.'

The device was simplicity itself and was exactly the same as the one Casey had found: a cigarette with three paper matches attached to it with a rubber band then concealed within a piece of yellow-lined notepaper. The paper hid the device when it was placed among highly-combustible foam products in the stores, and also helped the fires to accelerate once they had been lit. The beauty of it, says Lucero, was that 'this allowed [the arsonist] to time the device in relation to however those matches were placed on the burning cigarette. By screening the contents of the store he knew exactly where he wanted to place it. He could leave himself up to ten minutes to get away, depending on where he placed the matches in relation to the cigarette before all the devices ignited.'

That cigarettes are designed to continue to smoulder for some time after they're put down by the smoker was a help: 'He would take the match and he could place it on the cigarette and it all depends on the relationship to the burning end of the cigarette. If he wanted a short burn time he could place the matches closer to the burning end – that can give him anywhere from thirty seconds to one or two minutes; if he wanted a longer period of time to leave the area he could place the matches further down the cigarette, giving him five or six minutes and he could be on the freeway three or four miles away from the scene at the time.' This method 'allows him great flexibility, depending on whether or not he wants to be completely away from the area or if he wants to be a short distance away and watch the fire burn from its slow incipient stage all the way up to a large major fire. It's very brilliant in its simplicity.

'We recovered six devices during the course of our investigation,' Lucero continues. 'Every one ignited but what happened after the device ignited was partially his downfall. He couldn't tell if there was going to be a passer-by or an employee standing in the area of the device when it went off and was able to put it out or if the piece of material that he placed it in moved during the course of the fire and caused the device to extinguish earlier than he anticipated.'

From the beginning, the investigators were concerned for the employees and customers of the businesses involved – above all because the arsonist was setting fires that 'would grow in magnitude rapidly and with great velocity', as Lucero puts it. 'By his knowledge of fire behaviour and communication he was able to place these devices in strategic areas within the commercial occupancies to allow the maximum spread of fire in the shortest period of time. Through the combination of these factors we knew this was an especially dangerous individual.' Clearly, 'The person setting the fires, had some kind of knowledge about fires and fire behaviour. The average person really doesn't have a lot of knowledge about fire behaviour and communication.'

'It was like a game was being played with the arson investigators,' says Matassa.

As Lucero says, 'We felt the arsonist was setting those fires in a way that was almost taunting us or putting a red flag in front of our faces as investigators on these types of fires so we did want to catch the person.'

They soon had a nickname for him, as they often did for other serial arsonists, a name usually synonymous with the geography, the method or materials used to start the fires. In this case, says Lucero, 'he was setting a number of fires initially in a pillow or pillow material, and we called him the Pillow Pyro. The name stuck until we identified who the suspect was.'

As is usually the case when investigators are faced with serial arsonists, a task force was formed specifically to pursue the Pillow Pyro. Initially they were baffled and spread the story to other investigators around the state. When they came across Casey they offered to co-operate with him. Inevitably, the co-operation was 'unequal' – as Casey puts it, 'They said, "we're going to look in your files", and I shared what I had in my files. They did very little sharing of their files, but I shared all I had in my files, I don't think they had much in theirs.'

But Casey did hand over his precious fingerprint. The Los Angeles-based task force immediately had it tested. 'We got a call back from Ron George, the criminalist we were working with at the Sheriff's Department,' says Mike Matassa, 'and he says, "We got a hit – but it's one of your investigators." We said, "What you talking about?" He says, "John Orr. He should know enough not to handle the evidence. Tell that dummy not to handle the evidence." Everybody's jaw dropped with that remark because there's no reason John Orr should have been involved anywhere in an investigation in Bakersfield when he's in Glendale.' The prints of Fire Captain John Orr were on the database because twenty years earlier he had applied for a job in the Los Angeles police department – and had been turned down.

At first the investigators thought there had been some mistake for Orr was a fellow professional – and no ordinary one. He was in charge of protecting the wealthy Los Angeles suburb of Glendale and one of the best-known figures in fire investigation in the whole state, a man who had organized many seminars on arson and had personally instructed over 1,200 firefighters. 'I mean he was Mr Instructor, you know,' says Mike Matassa. 'He's interacting with everybody, particularly in southern California, in arson investigation and we were looking at him as a suspect – it was like we can't believe it. And in the 1980s fire investigators from round Los Angeles had formed themselves into an information-exchanging group which they called FIRST – Fire Investigators' Regional Strike Team. It's treasurer was John Orr. At the group's monthly meeting in March 1991 the retail store arsons were raised. No one had any comments on the case.'

Nevertheless, says Matassa, 'the younger people who really didn't know him or the people who didn't work much arson just wanted to go out and get him. Those of us, myself, my partner, Glen Lucero, who had worked with John in the past, knew that we'd better be darned sure before we go out and cuff this guy.'

Because his colleagues simply could not believe that Orr was the arsonist they asked Marvin Casey, casually, as Lucero puts it, if he had seen John Orr around the time he had recovered the fingerprint from the Craft Mart store. 'We felt at that time that he would say, "Yeah, John Orr was coming back from a conference in Fresno, saw the smoke from our fire, came over, offered assistance." As it turned out he said that John Orr had not shown up.' At the time Casey did not mention that Orr had been one of the ten investigators who had been at both the arson conferences during which fires had been set years earlier.

The only other investigator who wasn't surprised was Rich Edwards of the Los Angeles County Sheriff's Department. When he was told that a suspect had been identified he said simply, ' "It has to be John Orr", and he knew that by that point I was beginning to suspect that John was responsible for brush fires in portions of the county of Los Angeles.' These are difficult to investigate: 'We're dealing with vegetation, small bushes and trees and grass, and so it's very difficult to put that picture back to what it looked like before the fire – when these items are burned they become almost microscopic. Trying to determine what they are and looking for your fire patterns on rocks and in grass is very difficult and if a match or a device is used to start this brush fire, it's going to look like all the rest of the small pieces of grass and it becomes almost a microscopic square by square investigation, it's very difficult and can be very time-consuming if done properly.'

It was Orr's behaviour that had excited Edwards' suspicions. 'Over time John would arrive and quickly go to the area of origin. I found it a coincidence that in many of these fires John quickly recovered an incendiary device that he believed had started that fire. As an experienced investigator I knew how hard it was to do that and was initially impressed. As time went on that little voice spoke inside me, which comes with experience as an investigator, and the hair stood up on the back of my neck, and I just started to wonder. Initially I thought maybe he was just finding evidence for the purpose of being the hero but over time I knew that that couldn't be so, that he had to be responsible for some of these fires.'

All the investigators involved can remember the moment they discovered the name of the arsonist. At first, says Lucero, 'everybody was just in total disbelief. It was just not acceptable so we felt that we would double up our efforts to prove that it was not John Orr who was the arsonist. But the more we delved into the case and started working on it and trying to prove that he was innocent, more and more the finger of guilt pointed towards him as being the person responsible. It became a very sobering experience for all of us because when you're working on a big case there's a certain element of excitement, you enjoy the case, you enjoy hunting for the bad person. The ultimate goal is when you catch and identify the person and put 'em away. In this case all the fun element of looking for the bad guy was gone because it was one of our own and we knew it was going to send shock waves to the fire investigative community throughout the country.' Above all was the appalling feeling that 'one of our own had gone bad'. In Matassa's words, 'He's just a traitor, really, you know, to everybody else in the fire service, in law enforcement.

'Once we were sure that there was no innocent reason for John Orr's print to be on that device,' he goes on, 'we knew he was our suspect, but at that point we only had him on one fire and that's the Bakersfield fire where the device was recovered but there was not much damage. But we had twenty, thirty other fires that we're looking at as being tied into the same suspect so we had to channel our efforts towards gathering other evidence that'll help us tie him into these other fires.'

But they needed more than a single fingerprint. 'If we actually caught John in the act,' says Matassa, 'it would be icing on the cake. We would have him actually setting a fire while we had him under surveillance and there would be no defence, no way, no excuse for him to get out of being the arsonist.'

The chase was on. Orr, says Lucero, 'was out of his own city. He was on the road and it gave him multiple geographical areas which he could go in and go out. He's playing cat and mouse not with only one agency but with all the fire investigators in the state. In a way it was very strong to his ego to be able to show how much better and smarter he was than all these other investigators and then attend the conference with them. It was really him thumbing his nose. We knew that he was going back up to another conference, it matched the MO of his previous fires.' The conference was to be held a month later at San Luis Obispo on Highway 101.

'We decided,' says Lucero, 'that we would have to follow him to watch him twenty-four hours a day while he was attending the officer survival course in San Luis Obispo. Our thought was that we're not ready to arrest him for the fires that he has previously set but if there's the opportunity that he may set another fire during this seminar that we'll catch him and that would strengthen our case.

'But doing a surveillance on John was a challenge – I mean, this guy was incredible. He would take his police car, get on the freeway, get into the high-speed lane and just put the pedal to the metal. He would do over a hundred miles an hour on the freeway. If cars got in his way he'd put on his emergency equipment, clear the lane. To surveil him, you could not keep a car in sight of him because there's nobody else could possibly keep up to his speed.' And even if they could keep up, 'when you're travelling at that speed there shouldn't be anyone else behind you and to have another vehicle within eyeshot, he would easily determine that he had a surveillance team on him.'

Even surveillance from the air proved tricky. Normally, says Matassa, the aircraft or helicopters 'do circles because they don't want to go faster than the vehicle, but he was going so fast on Highway 101 they had to fly in a straight line and he was leaving us in the dust.'

Once they got to the conference, says Matassa, 'we wanted to be sure we had John covered at every moment so we had a surveillance team in a room across the hall from his room so that we would know whenever there was movement. We had surveillance vehicles surrounding the hotel, particularly paying attention to his car. We had foot units ready in case he decided to walk somewhere.'

'John would keep to himself,' says Lucero. 'He didn't seem to socialize in the evening with anyone.' Crucially, 'The first night there he went over to a drug store, walked in and came out with a pack of cigarettes. Everybody in the team says, "This is it, he's getting ready for action." We knew that the cigarettes were used in the device because he wasn't a smoker so everybody was geared up for something to happen. When you're doing twenty-four-hour surveillances you can get tired and your interest level can diminish a little. As soon as he purchased the cigarettes, the excitement level went up again.'

'We had heard that he was an adamant non-smoker,' says Matassa, 'but we wanted to be absolutely sure that he wasn't a closet smoker. We had agents that would be going through his trash to make sure he wasn't smoking in his bedroom at night. When he'd opened the door they'd walk down the hall, have a quick sniff of the room to see if they'd catch the smell of cigarette smoke.' Unfortunately some instinct of self-preservation ensured that Orr did not set any fires during or following the conference, for he had been warned of the task force closing in on him. After the fires in late March 1991, says Matassa, 'within a few days afterwards one of our investigators made a presentation at the task force that John belonged to. He described the device and the fact that there was a task force coming on it and from that point on to this date there has never been another fire that met that same MO. I think once he got the word that there was a task force out it was time to lie low and eventually we arrested him so there's still no fires to match that MO.'

Even before the conference they had taken video cameras and, as Edwards describes the operation, placed them 'high up on telephone poles in discreet areas and these would record vehicles coming and leaving the area over a twenty-four-hour period and if we would have a brush fire during that period we could pull the tape and look and see if we could see John or his vehicle. Another thing we did was place a tracking device called a teletrack inside his investigator's vehicle and we could track his activities at predetermined times as to where he was going throughout the county.' The results, though not damning, were positive. During the time Orr's vehicle was being monitored with the teletrack, 'We had two occasions just prior to his arrest where his vehicle was in the immediate area of where we had two brush fires.' Edwards firmly denies that these were coincidences: 'My experience as an investigator tells me that it's rare to have a coincidence. If we start having multiple coincidences then it's not a coincidence.'

Unfortunately, says Matassa, 'he caught our tracking device. We tried to cover it up when he found the device, the San Luis Obispo Police Department worked with us in trying to convince him that it was a hoax device but from that moment on he knew that he was a suspect and he knew that ATF and the task force was on to him. He never bought that it was a hoax device and that came out with some information we recovered when we finally placed search warrants on him later that year.'

To try to get further evidence they required up-to-date photographs of the suspect. This proved surprisingly difficult. Most of the photos were either out of date or of him in uniform, so, says Matassa, 'We finally came up with the idea of getting some help from the arson task force he belonged to. The president of that task force had been a former captain with the Los Angeles Fire Department arson unit so we approached him in confidence and asked for his help. He staged a PR shot, a sequence with the officers of that arson task force of which John Orr was one and that's how we got our photograph of him. Then we had to get some lookalikes. We went back to the arson squad and got together a mug line-up that was all arson investigators!'

The photo spread of Orr and his lookalikes, all pudgy, bespectacled, round-faced, proved invaluable. 'Several people started picking him out as being in the store that day, maybe a couple of times in the past. In fact one person at one of the stores said that they had seen him in here carrying some paper in his hand. That excited us because we knew that the paper that was used in the device was yellow notepad paper so we asked what colour it was. The response comes back, "Yellow." It was incredible that we're having this kind of luck with witnesses and it started falling together. Slow but sure, we were starting to put together the evidence to start bringing the case together on the fires beyond Bakersfield.'

The net finally closed around Orr after the single most extraordinary episode in a one-off investigation when the investigators came across the manuscript of a novel Orr had written. 'In the fall,' says Matassa, 'I'd learnt that John had written a book about an arson investigation and got a copy of a letter that he had written to a publisher. Incredibly in the letter it says, "My novel, *Points of Origin*, is a fact-based work that follows the pattern of an actual arsonist that has been setting serial fires in California over the past eight years. He has not been identified or apprehended and probably will not be in the near future. As in the real case, the arsonist in my novel is a firefighter." The arsonist in his novel is John Orr and it's incredible that he's writing this when he could have been helping us in the investigation.'

Once Matassa got a copy of the actual manuscript and started reading it, 'I couldn't believe what I'm reading. The fires we're investigating John for starting are chronicled in the novel. They're fictionalized, the names are changed, but the circumstances are identical, particularly when we get to the fires up in the San Joaquin valley.' Moreover, the many scattered arson attacks described in the novel could be known to only the person who had perpetrated them.

Even to an outsider the book is a strange, spooky read. The many long, detailed passages describing in detail how 'Aaron' sets the fires, why he chooses particular settings, and the work of the fire brigade in putting them out are convincing and exciting. The hero is a fire investigator who eventually tracks down Aaron, and after some nifty detective work in which he analyses the timings of the fires to see if they correspond to the times and dates in which certain shifts of firemen will be off-duty.

When they read the manuscript the investigators' hearts hardened at a description in the book of a fire that had occurred back in 1987 in a department store called Ole's in South Pasadena. The Ole fire was exceptional, because it was the only one in which people had died. The victims, says Edwards, 'were trapped. They knew that their original exits had been blocked. They knew those doors had come down, the lights are going out, the smoke is filling, it's black, we have a grandmother with a grandchild who's attempting to flee. And in John's manuscript Aaron describes that the reason these people died is because of their own stupidity. It's not the arsonist's fault, they died because they were stupid. I just find that chilling and, no matter what the motivation is, for somebody in the professional fire or police service to be involved in that kind of fire is very troubling for me.' As one of Orr's characters remarks, 'cops go bad all the time, but not firemen'.

The whole incident, says Lucero, 'just drove home how dangerous these fires were and the potential life hazard for any of the fires that he set in the past or any fires that he may set in the future. There was no doubt in our minds that since we discovered in 1991 that he was setting fires back in 1987 in commercial occupancies, that he was going to continue setting fires. We knew that he was callous because he had already killed four people at the Ole's fire and had continued to set the same type of fires. It showed him to be very calculating, cold-blooded and with no mercy for anybody. We wanted to catch him all the more but we wanted to have a strong case against him too. That's why we didn't arrest him initially.'

'When it came time for John's arrest,' says Edwards, 'the federal task force determined that he should be confronted by people that he knew when he came out of his home. They felt that that would confirm to him that we knew that he was responsible for these fires. The other reason for that was for safety purposes. They felt that if he recognized investigators that there would be less of a chance for a violent arrest. When John came out of his home to walk to his car he was confronted by investigators that he had known over several years, that he'd worked with. He seemed surprised and offered no resistance.'

'When we arrested John,' recalls Matassa, 'he was carrying a briefcase. When we opened it we found the same brand of cigarettes that were used in the device, matches, the rubber bands, all the materials. Then when we searched his car, underneath the floor mat we found the yellow paper of the same size that was used in the device. Everything that was needed for the devices that were recovered in the investigation were in his possession at the time of his arrest.' And when they searched his house they found a macabre video library of devastating fires stretching back over a decade, to a time well before Orr's activities had aroused Marvin Casey's suspicions. The videos showed many fires in their early stages, which led the investigators to believe that Orr had set most of them.

'After we arrested him we sat down and talked to him. At one point we got to a conversation about the fingerprint. I told John that we had found his fingerprint on a device and his response was a question. He wanted to know how many points of comparison I had and when I told him that it was irrefutable that it was his print, his next question was, "Well, how many prints did you get?" I mean, an innocent man would have been telling me you couldn't have my print, not wanting to know how good a print it was and how many I had. It's not an admission but it sure as heck isn't the response of an innocent party.'

Originally Orr was found guilty of six counts of arson, including the two fires back in 1987 that had originally aroused Marvin Casey's suspicions. For them he was sentenced to thirty years in jail. But, thanks to the task force's efforts, he was also found guilty of a further twenty-three counts of arson and four counts of murder and sentenced to prison for life, with no possibility of parole. Orr has never admitted his guilt, or talked about his crimes.

This left the investigators to try to work out his motives. These were not necessarily the same as those of his fictional doppelganger, even though one of the investigators in the book explains the nature of a serial arsonist in terms that suggest it resembles Orr's analysis of his own behaviour. 'The serial thing,' explains the investigator, 'usually starts after they have experimented with fire when they're young, and just continue it if they aren't caught early. As they grow up, it takes on a sexual atmosphere. You know, they are too insecure to relate to people in a direct, person-to-person way and the fire becomes their friend, mentor and sometimes their lover. Actually a sexual thing.'

The 'Ole' passage also provides a number of clues as to Orr's own train of thought.

The fire was ineptly termed accidental [the true professional coming out!]. Aaron was so furious that he set a nearly identical fire two days later in Hollywood at another hardware store. The investigating agency termed the fire arson but no correlation was made to the Cal's [i.e. Ole's] fire. Aaron wanted the Cal's fire to be called arson. He loved the inadvertent [sic] attention he derived from the newspaper coverage and hated it when he wasn't properly 'recognized'. The deaths were blotted out of his mind. It wasn't his fault. Just stupid people acting as stupid people do.

Matassa naturally got 'that sinking feeling that what we had hoped we had prevented had happened before we had a chance to prevent it. Innocent people had died because of John's fires. What really strikes home is how cruel and uncaring he is.' Particularly sickening was Aaron's remark after a baby had been killed in one of his fires: 'It was too bad about the baby but, shit, it wasn't my fault.'

'It's like the two sides of John,' continues Matassa. 'There's the arson investigator who's the hero, macho man, everybody looks up to him and yet there's the arsonist firefighter who is a loner, a despicable character that nobody wants to have anything to do with and it's like it's both sides of the John Orr we're investigating.'

To Matassa, as to his colleagues, 'This investigation was probably one of the high points and the low points of my career. The low point was that a fellow investigator was the arsonist and people lost their lives because of him but the high point was the challenge of the investigation, of tying all of the little bits of evidence together, and the fact that once we got him on the initial fires we were able to develop the investigation that led to him being prosecuted for the murders of the people at the hardware store. Satisfaction as an investigator, that's about as good as you're going to get.'

Which leaves the great question that Orr himself refuses to answer: his motives. To this day, says Glen Lucero, 'Nobody can find an answer for what caused somebody who had such a fine career going for him, and had so much to offer, to go so bad.'

Rich Edwards believes that he had 'multiple motivations. They may even have been monetary. We have indications that John set fires or volunteered to set fires for people for insurance purposes.' Nevertheless, for Edwards it was the novel that provided most of the clues as to Orr's motivation: 'When I read John's manuscript,' he says, 'I was struck by the fact that much of it was written from first-hand knowledge and he may have been driven to set some of these fires so that he could describe them in his book. Some of the fires he could have learnt about through his association with other investigators but I was struck that whoever wrote this manuscript experienced portions of what was described in it. There was a real-life description of things that he had done, not hearing about the fire but setting the fire and sitting back as Aaron and watching it burn and feeling the satisfaction that derives from that. I think John enjoyed the deception and the power, and when you're setting fires and watching other people respond to that, whether it's the victims running from their home or the Fire Department responding, that's power, you are controlling the circumstances.

'By setting these fires he controlled the responses of the people who were victims of the fires, he controlled the responses of the firefighters and he controlled how the investigators would deal with fires. As time went on – from the manuscript – if these fires were not investigated properly or were undermined, John described how he became angry because they didn't call the fire correctly, maybe they called it undetermined when it was in fact an arson.'

'What I would really like to see out of this case,' says Glen Lucero, 'is for John Orr to take his responsibility seriously. Since he's got four life sentences to face, he should tell the truth, tell us why he did it, what caused a person with a high position in the fire community to commit these crimes. Maybe we could learn from it how to prevent somebody else from going down that path in the future.'

'I don't know if John is capable of sharing his inner thoughts with us,' says Rich Edwards, 'but if he could it would be a great learning tool for other fire investigators. John has always been a champion of investigators learning from other investigators' experiences so I would hope at some point he would open up. But after many years in the criminal justice system, people defending their case get to a point where they believe their own lies, and the more time that goes by the less likely it is that he will be forthcoming with any legitimate information.'

It's not just the investigators themselves who need replies, says Lucero, he has an obligation to explain himself to the families of the people that he's impacted by these fires, the civilians, the employees of stores that burned down so they had no jobs, and the relatives of those killed in the Ole's fire. It's for them that I would like him to come forward and explain why he did it.'

After reading Orr's book, my own belief is that, although early in the book Aaron boasts that 'No one will ever catch me' Orr expected to be found out and that he was resigned to dying in the process of discovery. The novel ends when, in the best tradition of all such thrillers, the hero catches up with Aaron/Orr, in an empty store he has just set alight, and kills him in self-defence. It looks as though Orr was preparing himself for a sort of Götterdämmerung, in which his death would be the last in his increasingly desperate search for more and greater thrills.

# IV

# THE HUMAN FACTOR

# 11
# We're Only Human

You must think about people's reactions to a fire in terms of the three basic stages of making sense of what's going on, preparing to act and then acting.

Professor David Canter, Liverpool University

How we react to fire depends on how much information we have about it. We need answers to three important questions: Where is the fire? How severe is it? Should I stay or go? When things go wrong it's generally because people didn't know what was happening and were unable to take the appropriate actions to save themselves.

'In the initial stages,' says Neil Townsend,* 'people behave in fires in exactly the same way as they behave in normal life. They start getting some kind of signal, a stimulus or whatever, and they react to that. Part of the problem is, in the initial stages of a fire people would expect to encounter heat, smoke and flames, but they'll encounter some different signal, maybe a window breaking, a noise or strange smell they can't understand. They'll either ignore it and hope it goes away or they'll investigate. I think in the initial stages people just behave how they normally behave anyway. Once they discover it's a fire a different kind of behaviour takes over.'

In a surprising number of cases the system for warning those at risk is either lacking or severely defective, even in the largest, most complex and vulnerable structures. Nowhere was this better illustrated than in the MGM Grand Hotel fire in Las Vegas as recently as 1980. No alarm was sounded in the thirty-storey skyscraper that housed the hotel's bedrooms. The fire

---

*A divisional officer in the London Fire Brigade who says that his job is trying to 'amalgamate knowledge and look at the fires in London where people are involved'.

141

was burning many floors below in the casino and it seemed ridiculous to suppose that people were in danger so far away.

When lethal smoke started spreading throughout the tower, hotel residents were alerted to the danger they were in without knowing where the smoke was coming from. No other warning had been given and the first many knew of the fire was when the other hotel guests went down the corridors banging on doors, yelling, 'Fire.'

According to Steve Hanson, deputy fire chief at the Clark County Fire Department, 'Because of the time of day they were either still in bed or just in the stages of getting up, getting dressed, getting ready to go out.'

'There was a [legally adequate] fire alarm system in place,' says fire analyst Carl Duncan. 'The most effective means of notification in this particular fire was security and hotel employees notifying each and every guest-room floor. There was no voice communication to the rooms at that time and no way for occupants of rooms to be communicated with other than by telephone. Many tried to call the front desk and switchboard operators tried to ring rooms to notify occupants but the intensity and speed of the fire development soon rendered the switchboard areas untenable.' The system, he adds, in a masterpiece of understatement, 'was not a comprehensive, well-designed fire-protection system.'

Because, says battalion chief Ralph Dinsman, 'there were no announcements to any of the occupants on the upper floors of the hotel it was natural for the people when they heard commotion or smelt smoke to go out in their hallways and once they saw the smoke the first thought that comes into their mind is to get out of the building. So they would proceed down the hall to the stairways which had more smoke in them than the hallways. Knowing that they couldn't get down or go up they came back into the hallways but because they had left their room key in their room they couldn't get back into their rooms. Tragically a lot of people perished in the hallway right outside their room, where they should have stayed. Because most of these hotels are made out of steel and concrete, they are pretty much fireproof and the doors are fire-rated. As long as you follow instructions with towels under the door and things like that, you're in a pretty safe environment.'

The smoke must have been truly awful. Even when Hanson and the other rescuers got there much later and searched the lower floors of the complex 'which didn't have windows or doors open to the outside, the smoke was just as thick as mud and you had to get down and crawl around. On the upper floors it would have been easy for them to become disorientated and if they couldn't get back into the room because they didn't have their key or they couldn't find an exit, that's where they died. The smoke would have been burning their eyes, burning their throat, making it difficult to breathe and in a short time if you didn't get down to the ground you would fall to the ground overcome by it.'

'People staying in hotel rooms,' says Dinsman, 'sometimes know what to do, sometimes use common sense, but other times they may panic if they see smoke or see fire. At the time of this fire a lot of people didn't have instruction and when they looked out their doors or heard the alarms or heard the commotion, it was natural for them to leave their room and try to exit the hotel. That was one big mistake that a lot of them made. If they had taken precautions, put wet towels under their doors and stayed in their rooms they'd probably be alive today.'

Clearly there was chaos, starting in the casino where, at 7 a.m., there were only a few hardened gamblers. 'There really was not an effective fire-emergency plan,' says Carl Duncan. 'There were minimal staff, pre-dominantly security and, of course, the gaming-table staff and the pit crews. They took the time, believe it or not, to cover up the gaming tables, secure the funds and make safe egress while encouraging occupants to do the same, but there was no concerted effort on the part of hotel staff to form a comprehensive and effective evacuation plan for the guests.'

The gamblers' reactions varied widely. Some, says Duncan, 'were informed and perceived the risk so made immediate exit'. But there was one gambler 'who was wanting the dealer to deal one more hand of cards before he got up and left the table with his winnings or without his winnings. We had witnesses that observed patrons getting angry at the croupiers for locking up the gaming table and asking everyone to exit the building. We had one incident where a patron was found down the street gambling at another casino with chips he had stolen during the course of his exiting the hotel that was on fire. Security was admirably attempting to get everyone out and in fact several security guards lost their lives at the casino level because they tried to assist.'

'We learned,' said Dinsman, 'that communications were poor, with the other agencies and with the hotel guests. We learnt a lot of things from this fire and our legislaters took heed of this. Months later we passed legislation to install sprinkler systems, to have fire-control rooms, to have audible alarms and information that would warn guests of any fire or danger in a hotel.'

Three months after the disaster at the MGM Grand there was another fire in Las Vegas, this time at the Hilton Hotel. The fire started on the eighth floor of the hotel in an elevator lobby, but 'only' eight people died. 'We were able to take a lot of the lessons that we had learnt on the MGM,' says Ralph Dinsman, 'and use them on the Hilton fire. One was helicopter rescue from the roof, we did that again on this fire, and another one was getting information to the occupants of the hotel. We were able to do this because the local news channels had their live-eye cameras on the scene and I was able to go on live television, and have the hotel notify the guests to turn on a certain television station because the Fire Department would be talking to them and telling them about safety measures.

'I was able to talk to these people and tell them to stay in their rooms, don't panic, the Fire Department's on the way, they'll be there to rescue you but the safest place to be right now is in your room. Wet a towel and put it under the door because there's usually a crack where smoke can get in. Don't open your windows if they're sliders because smoke may be sucked into the room. Go into the bathroom, draw some water into your bathtub so that you'll have a source of water in case the fire was right in your hallway and you needed to extinguish the door or keep it cool. All these instructions were to have them feel safer and know that the Fire Department was there. Later we were able to let them know that we had the fire under control and other information that was pertinent.'

The new system worked. Dinsman remembers, 'Some people came down after the fire and said that they watched the television and they did exactly as they were told and stayed in their rooms. It saved switchboard operators having to call every room in a hotel with over two thousand.'

The effect of the fire on public perception of the risk in staying at a large hotel-casino was enormous. In Tom Klem's view, (a fire protection engineer) 'The MGM blaze focused on the potential hazard of high-rise buildings and the potential for a high-rise building to result in a multiple-death fire. As a result, the public focused on what they could do, what kinds of actions or reactions were appropriate in high-rise buildings. I think that the people who viewed the film footage of the MGM fire realized that it could have been them trapped in that high-rise building so they asked very pertinent questions. What would I have done? What should be done? How could or should I have reacted to that particular fire scenario?'

As a result 'a wealth of information began to emerge from the fire-protection community with regard to how you can survive in a high-rise building. Do I go out into the corridor? What happens if I am confronted by smoke? If I'm confronted by smoke is there a way that I can safely ride out the fire by having towels moistened and placed at the bottom of the door or placed over my face? Should I open or close the window? Exactly what *do* I do? A whole wealth of information about survival techniques in high-rise buildings, and in particular hotel occupancies, was brought to bear soon after the MGM Grand fire.'

What the MGM Grand fire, and countless others, demonstrate with painful clarity is the need for complete, comprehensible, reliable and unmistakable information to be disseminated immediately after a fire has broken out, be it in a hotel complex or in a private home. As Neil Townsend, divisional officer commander with the London Fire Brigade, says, 'If you have a smoke alarm in your home that starts sending out a signal then you have relatively reliable information coming in that there's a fire or at least that something's made that smoke alarm go off and it needs checking out.

'The first information people get about a fire is probably wrong. Something else may be happening. Perhaps there's a bonfire or a fire outside or a car alarm is going off. I think this is due to the fact that when people think about fire they think of flames, heat and smoke so if there's an unusual noise or an unusual smell they don't tend to associate that with a fire and they need to investigate a bit further. Everyone I speak to generally makes that same mistake, that they misinterpret the first signals, the first information they're getting.

'Getting people out of a fire depends on making them realize that this is an emergency, that they must stop what they are doing and react to save themselves. The speed of our reaction to a fire alarm will depend on how committed we are to finishing the action in which we are already engaged. If we find it difficult to do this when the fire is burning before our eyes, how can we be expected to change gear when the fire may be far away and we can't even see it? When the only information that we have is the smell of smoke, or an alarm bell ringing, or a vague rumour that there's a fire somewhere in the building, we have a problem.'

Above all Townsend emphasizes the ambiguity of the first information we receive about a fire. 'Even instant alarms have a limited value. Traditional regulations assume that people will start to leave a burning building the minute they hear an alarm, but the ambiguous nature of many of the first clues to a fire and the way we interpret them makes this a false assumption. When we are alerted to a fire we tend to behave in a predictable way: we go and investigate further. This is actually a sensible thing to do. As Jonathan Sime, an environmental psychologist, says, if every time we heard an alarm go off or smelt smoke we stopped what we were doing and left the building we'd never get anything done. We are faced with false alarms all the time. 'I investigated a fire when a school deputy head set off the alarm for a fire drill but then had to run round the school telling the teachers to evacuate the pupils because it was a real fire.

'When we get information that something, we don't know what yet, is going on, we try to match up the clues with a known event that we have already experienced and that's very rarely a fire. Our perception of the event is coloured by what we already know. We're not very good at knowing what to do in completely new situations. And once we've worked out what is happening, our reaction to it will depend on our attitude to that event. The attitude is affected by many different factors, including the way people around us are behaving and how dangerous we perceive the situation to be. Only after all these processes can we start to react. In a fire, seconds count and the faster you realize that you're in danger the greater your chances of survival will be.'

The behaviour of a serious fire creates its own problems says Professor David Canter of Liverpool University, 'If you watch the growth of a fire it usually begins relatively slowly, then takes off quite quickly and eventually can explode. People have difficulty in understanding that what they see as a perfectly manageable fire in the early stages will grow out of control very quickly. They think it's going to grow steadily and will act in the same way at the later stages as it did at the early stages, but the fire will take off and they will no longer have the possibility for escape or for managing the fire. People just don't understand the physics of fire growth if they haven't got any training.'

Only too often the very people who should be issuing the warnings create problems by ignoring these rhythms. As Professor William Feinberg, of the University of Cincinnati, says, 'Those who are responsible for notifying people about evacuation often delay making the announcement because they're concerned to avoid panic. We have this idea that panic will kill, but the best action to take is to tell people quickly and frequently with explicit directions about where to go, how to exit. This is very important because time is the big enemy to successful evacuation. The sooner it gets started the better off everybody's going to be. In the Beverly Hills Supper Club fire [see Chapter 13] this long time delay is what helped kill people as well as there not being enough exits. Had the guests been warned sooner, more people would have evacuated successfully. So this whole myth of panic, the concern not to tell people because that might cause undesired behaviour, doesn't seem to be the thing that we need to worry about. What we need to worry about is delay.'

If the first lesson for escaping from a fire is to be able to receive relevant information, the second is how to interpret it correctly. The official report on the Woolworths fire (Chapter 6), in which the initial warning was given verbally by staff, describes how 'the fire alarm was not actuated in the initial stages of the evacuation. Several people claim not to have heard any alarm signal, some stated that they heard bells ringing after they had left the building, and some claimed to have heard a bell which sounded for about a minute but did not resemble a fire alarm.'

The next step is how to interpret the information you have received, and, a more difficult problem, to be persuaded that you should act on it. Between information and action come a wide variety of reactions. 'There's a great number of factors responsible for why we have such a high proportion of people dying in their own homes,' says Neil Townsend. 'Some are alcohol-related where they probably knew nothing about what was going on anyway.' But even when they're not fuddled 'people tend to misinterpret the first fire signals that they're getting in favour of something else and will either ignore them and hope they go away or investigate and become involved in the fire.

'When people are first alerted to the fact there's a fire they don't immediately interpret the situation as a fire. It's a strange smell or strange noise while they're lying in bed at 3 o'clock in the morning and they're not quite sure. An unfolding drama is going on behind the scenes, as it were, and they're not quire sure what's going on and eventually they go and investigate. So if we had a fire in a lounge the parents come downstairs to investigate. They open the front door or the door to the lounge and, suddenly confronted with heat, smoke and flames, they know it's a fire. Now, part of the problem is they have left the children upstairs while they investigate and now the children are trapped.'

The older you are, the more you are at risk. 'The majority of people who die in fires at home,' says Townsend, 'are over sixty-five. It's more likely that someone of that age is going to be single too, so you can imagine a single sixty-five-year-old person in bed at 3 o'clock in the morning and hearing a window break downstairs or wherever . . . If they come to the conclusion that it's a burglar trying to break in I'm pretty sure that most of them would hide under the covers and hope he goes away. This means that they've lost a lot of valuable escape time.

'I think they pick up on the fact that there's a fire once they start getting smoke into the bedroom and by then probably their escape route is cut off by a lot of heat and smoke and it's too late. They either need rescuing or they don't survive. I've been to a number of fires where the last thing anyone has done is open the door to the lounge and been consumed by the fire. If there is anyone left upstairs, maybe children, their escape route is now cut off because the flames, heat and smoke go rushing up the staircase and they need rescuing by the fire brigade.'

The sheer speed at which a fire can take hold combined with the natural reluctance to abandon an enjoyable treat was a major factor in the fire at the MGM Grand. As firefighter Steve Hanson points out, 'They thought they were in a safe environment. They were doing things that they flew or drove miles to do and they just didn't want to leave or accept that they needed to get out of there. It's just not something that was on their schedule. 'They were in Las Vegas to have a good time. They were in a good environment, having fun, getting free drinks, gambling, winning. The last thing on their mind was a fire or an emergency where they had to get out.'

The same type of reaction proved fatal in the Woolworths fire. As Stan Ames, fire consultant pointed out, 'The tests not only showed how fast the fire grew but also showed us how toxic the smoke was. We knew that if people inhaled the smoke they would probably have become unconscious after only a few breaths. The other thing that surprised us was that although people could see the fire growing they didn't react to it immediately. Since then our studies of human behaviour in fire have moved on but at the time we didn't realize that people who were sitting down to lunch were far less likely to move than people who were shopping.'

'I've been involved in fire drills in large department stores,' says Neil

Townsend, 'and I've been in department stores when the fire alarm's gone off. It's interesting to watch people who have spent a long time going round the store filling up their trolley. They're reluctant to leave it, even though they haven't paid for it. They've invested a lot of time and effort in filling this trolley and they're very reluctant to leave it on the sound of a bell that could even be a burglar alarm.'

At Woolworths the reluctance to move was greatest in the store's restaurant, the epicentre of relaxation. 'Initially,' says Steve Wood, a survivor of the fire, 'people in the restaurant area didn't react until they saw things start to happen, the smoke appearing, the flames. That was when panic started.'

'People were sitting down having their lunch,' says Stan Ames, 'they saw the fire start, the staff became aware there was a fire and they asked people to leave. A lot of people refused to move when they were asked. They were sitting down having their lunch and they felt it wasn't necessary for them to move so quickly.'

'The first few people I spoke to,' says Steve Wood, 'thought it was a fire drill until I pointed across the sales floor and you could see the flames through the glass in the side of the restaurant. I went round the restaurant, asked a few people to leave. One chap said, "I've waited long enough for this meal, can I finish it?" I said no, he'd better go.'

But lessons were learnt from the Woolworths blaze. 'We've been able to use this information,' says Stan Ames, 'in our understanding of people's behaviour in fire, in the planning of fire-escape routes and protection measures, and we now realize that not everybody will behave the same when a fire occurs. There are certain activities people are engaged in that may prevent them from leaving immediately and this was a classic case. I think it's been cited time and time again in fire-safety literature.'

# 12
# The Fatal Power of Normality

If we study what actually happens in fire situations, if you look at King's Cross people were determined to go up the escalator and try and get out of that building in their familiar route.

Neil Townsend, divisional officer, London Fire Brigade

When an alarm goes off, or smoke gets into a room we don't immediately jump up or rush straight for the fire exits. Psychologist David Canter points out that human beings have a benevolent attitude towards fire: that we associate it with positive things, warmth, food, fun, and that we control fire and have been able to do so since prehistoric times. To most of us, virtually all the time, fire is not threatening. This is of overwhelming importance when we are faced with a fire emergency.

Eventually the clues stop being ambiguous and we are forced to react to a fire. We reach a point where we realize that the situation is an emergency and we must act to save ourselves. What we do next throws up distinct patterns of behaviour. One of the most important factors in understanding why people behave as they do is related to the psychologists' theory of the 'script' we follow, which itself centres on what we feel our role should be in a particular social environment.

It takes a lot to get people to change 'the script' of the behaviour patterns they find normal in a particular environment, be it at home, in a hotel, office, restaurant, department store or at a football match, when they are confronted with even the most unambiguous evidence of a fire. Because we know what the psychologists call the appropriate 'script' so well, having to break from the normal pattern is difficult. This is one of the keys to understanding how people behave when they are caught in a fire.

One of the most consistently demonstrated patterns of human behaviour in fires has been that people do not like deviating from their usual route in or out of a building. In a fire emergency people often try to escape by the same way they came in: when questioned why they didn't use a fire exit that was closer, they often reply that it had never occurred to them to do so.

'Obviously,' says Neil Townsend of the London Fire Brigade, 'we label fire exits and alternative ways out for people but if you're caught in a fire you're hardly likely to look about for the nearest exit. Most people leave a building by the way they're familiar with. There's been a number of occasions where a fire has occurred and people could have got out the window of their office but chose to leave by the front door and in doing so have not made it. But people are comfortable with what they're familiar with.'

For Professor David Canter, the most remarkable thing about people's behaviour in a fire is that 'they carry on with their ordinary, conventional, day-to-day activity with the script that guides what they do when there's no emergency. They follow through on that until the circumstances are so dramatic, so disturbing, so demanding that they feel they have to do something very different.

'We all know what we're expected to do in a restaurant,' says Canter, 'the sequence of activities. It is a bit like a script for a play. You come in, you sit down, you get the menu, you order, you wait for your order to come and you pay for your order at the end. These stages are expected and laid out for us in conventional ways and anything that happens around that activity is absorbed into the script. So initially if you heard some noise that you thought was coming from the kitchen, say, you would think that that was just something to do with the day-to-day activities of the kitchen and even if you heard shouting you might interpret it differently and you might think that it's some sort of fight or row or something, you wouldn't think it necessarily relates to you because normally what goes on in a restaurant kitchen doesn't relate to you at all, so you would carry on with your normal pattern of activities, misinterpreting the indications that there was something going wrong and, of course, that's terribly dangerous.'

Equally, if the fire is your own house, Canter continues, 'You may well deal with what would normally happen in a domestic situation. Normally it would be expected that the man would go and deal with the fire or the woman would help get people out, and it's interesting that the men, therefore, are more likely to be injured because of that but it's because of the existing pattern of relationships.' In a department store, say, there is also 'an expected set of relationships': if they 'are buying things from the counter they may feel that they ought to pay for them so they may wait.'

Researchers found that at the fire at King's Cross Underground station, described in Chapter 3, people were unwilling to deviate from their usual paths through the station. Paul Godier, head of safety and environmental development of London Transport says, 'There were examples of customers on their way home, wanting to catch a train upstairs in the main-line station who, when told by staff not to go that way just brushed past them. There was even a point when there were tapes across the escalators yet people just unhooked the tapes and got on to the escalators. When people have a firm intention to do something, it takes a lot to get them to do it differently.'

After the fire, says David Canter, the authorities provided enough information to trace the pattern of how the victims moved. 'We were able to establish where the bodies of the victims were found, then to look at the police records and find out from that where those people were likely to have been travelling and the direction they were likely to have been going in. Many people cope with travelling on the London Underground as if they're blindfolded, just going in a direction they know without thinking. In the emergency, when it was filled with smoke, what they did was they just went straight through in more or less a straight line, not able to see what was going on because by then the smoke had made it so dark and they just kept on going in a straight route, out to the back and some of them unfortunately were overcome by the smoke and died in that location.'

The researchers, says Canter, found that the victims 'were not scattered around randomly in the way you might have thought, with people sort of overcome by flames and smoke and the whole panic of the situation. Most of them were moving through in exactly the route you would have expected of an ordinary traveller's journey. The people who had been going out towards Euston Road were the ones who were found dead over in that area, the people who had been moving out towards St Pancras, those were the individuals who were found dead over there. It was quite remarkable the way in which this showed us that people in these remarkably threatening and dangerous circumstances continue with their day-to-day activities in the way we do if it isn't dangerous because that's all we know to do.'

'People stick to a script,' says Neil Townsend. After looking at a video of what happened at King's Cross, he saw how, 'they come off this train, walk up the platform, up a particular escalator and out. There's other videos where no one is prepared to make a fool of themselves and now this guy as he comes in knows there's a fire, he's looking at it, he joins the queue behind a woman who already knows there's a fire because she's seen it but they're too embarrassed to say anything. There's this sort of mutual ignorance going on. Eventually when one of them said something, the other one jumped in at exactly the same time.'

Possibly the most dramatic recent instance of a mass refusal to break the script came with a fire at a football stadium. In May 1985 Valley Parade Stadium in Bradford was the venue for the last match of the season for third division Bradford City, headed for promotion to the second division. Naturally there was a capacity crowd to watch the game against Lincoln City. When a small fire started under one of the stands, the people closest to it reacted with a mixture of curiosity and amusement. One fan, David Pullan, saw 'a plume of smoke and nobody thought anything of it. Prior to the game fireworks had been thrown and perhaps this is what people thought it was.' The first thing that Sergeant Glynn Leesing, one of the policemen at the match, noticed 'was a smell of burning. This was the first time we'd actually worn some new coats that they'd issued us and they were made of plastic. Initially I thought somebody was burning the back of my coat with a cigarette.' Some of the spectators even took photographs as they waited for the police to bring a fire extinguisher from where they were kept in the club room. Other spectators in the stand took barely any notice. They were there to watch the game and weren't going to be distracted by a bit of rubbish catching fire – they were determined to 'finish their own script'.

But a short time later some of these spectators were dead in their seats: they hadn't had time to budge before the fire took hold. Within minutes the structure of the seventy-year-old wooden stand had caused the fire to flash-over. Smoke and flames poured down the length of it faster than people could move. When David Pullan and people round him decided to leave, 'We made our way to the back of the stand because that is the way we'd come in and we got to one of the exits. They were only wide enough for one person to get through at a time. There were perhaps twenty people milling round this exit and nothing was moving, obviously because the passage at the back of the stand was already full of people and the back exits were locked. I think I first started to realize that things were getting dangerous after I'd been there for perhaps thirty seconds and nobody was getting out through that exit. I looked up towards my right and I could see a wall of flame shooting down the back of the stand. At that point I realized that we weren't going to get out there so I made my feelings known to the other people around us that, you know, we weren't going to get out there and I repeated this a couple of times. Nobody else moved but I decided that I was going to move so I started making my way down to the front of the stand to get over the wall there.'

But the others were so psychologically conditioned to getting out only through the exits and not by the pitch that, as Pullan says, 'They didn't move with me. I never saw anybody after that. It's funny, I don't know if it was the case or not but as I were making my way down the stand I felt as if I was entirely on my own, like there was nobody else around me. I got down to the perimeter wall, which was about four foot high, and I put my hands on it to lever myself up and a youth on the other side of the wall just grabbed hold of me and dragged me over.' The whole thing had taken a mere five minutes.

'Some of the spectators,' says Sergeant Leesing, 'were quite bolshie. It was as if we were spoiling the day. "You deal with the fire or the smoke, I'm happy." It was terrible. They couldn't see why we were telling them to move. Had I not been a police officer and a steward, I don't think people would have moved.' Even then, he says, 'they wouldn't go on to the pitch. Their natural reaction was to go out the way they came in. They wouldn't climb over the fence. Everybody was trying to get out of one small exit and I would say that within a minute the part of the stand where the whiffs of smoke had started was well alight, seats, part of the roof, and people were still jammed in the back exit. Joe [a fellow policeman] and I went to the top to the exit and I grabbed hold of a woman but she wouldn't come backwards where the fire was. She was going out the way she'd come in.' At that point the police simply had to save themselves.

And why did the spectators not try to save themselves? 'Apart from there being a four- or five-foot wall in front of them,' says Sergeant Leesing, 'in any other game it would have been taboo to climb over the wall and go on to the pitch. You would have got arrested. It's just not a natural thing to do.' After the fire had been extinguished fifty bodies were found. Many more people had been badly burned and were in hospital. It seemed absurd that so many could have been trapped in a building that was open at the front. But the Bradford supporters had not realized just how dangerous the fire was. They assumed that if the worst came to the worst they would have plenty of time to make their escape by the way they normally left the stand.

Another major complication is the role played by the staff on a fire in a commercial building. At the Summerland leisure centre fire on the Isle of Man in 1973 and at the Beverly Hills Supper Club in Kentucky restaurant and bar staff thought that they would be able to contain the fire and tried to tackle it, not wanting to cause disruption and alarm when they didn't think it necessary. Raising the alarm was delayed.

But, as David Canter points out, everyone continues to play their role from their pre-ordained scripts, sometimes until it is too late, or until someone gives a warning credible enough to result in action. At the Beverly Hills Supper Club fire, Joe Swartz, an NFPA investigator, found that as the fire spread through the building, the nightclub staff continued to do their job and look after the patrons. He saw that they consistently acted in 'other-serving' ways, while the patrons acted in 'self-serving' ways. This is consistent with the roles the two groups had constructed for themselves: the patrons were there to be looked after, the staff to take care of their needs. One guest that night was a local firefighter. First of all he made sure that he got his family to safety, remaining in his role as husband and father; once he'd done his duty by his family he switched to his professional persona and returned to help rescue other guests.

'Generally,' says Professor Norris Johnson, of the University of Cincinnati, 'there was a distinct difference between staff and patrons. There was what some people have called the maintenance of their roles. The first concern of most employees was to help their specific customers out. There were any number of occasions in which servers went back to the area they were serving and took care of those customers they were responsible for, two or three referred to people who were regular customers so they were concerned about assisting them. When you recognize that there's a clear difference between what employees do and what patrons do, between what men do and what women do, you see that all these expectations of how people behave in normal situations seem to continue even in this emergency situation.'

But, crucially, the credibility of a warning demands that the listeners accept the authority of the person delivering it. In some environments, says David Canter, there's a natural hierarchy. 'If you're caught in a fire in a hospital it's relatively safe because you expect to do as the nurses tell you. They are part of a hierarchy, of a military structure, really, and they are set up to deal with emergencies. Any self-respecting patient does as they're told. If the nurses tell you to get out, you get out, and so it's actually quite safe because of that organizational system that will guide you through. But there are other circumstances in which the organization is the wrong way round. In a hotel the staff don't normally tell guests what to do, the guests tell the staff what to do, so if there is an emergency the staff are reluctant to tell people to get out. It's not part of their day-to-day dealings with guests to give them those sorts of instructions.'

To make matters worse, customers in any situation are unprepared to take any responsibility. They assume, says Townsend, 'that it's the problem of the staff, because they're a customer and the staff always help them. This has been demonstrated in the Woolworths fire and the Beverly Hills Supper Club fire where customers want to be directed by the staff, even in times of emergency like this. It's their problem, it's not our problem, we're just customers.'

'In the King's Cross fire,' says David Canter, 'there were different officials, police officers, station staff, their supervisors, engineers, a whole variety of people, as well as the passengers moving through the tube station, and they all had relationships that existed before the fire occurred. Those carried on in the emergency so that people did things depending upon what other people told them to do, what they thought was required of them to do and what they understood about the whole pattern of activities. For instance, police officers assume that they would have to take control in an emergency and start giving instructions, members of the public assume that if there is a police officer around he will tell them what to do whereas station staff assume that they need information from a supervisor unless they've got instructions to tell them to operate differently. So that whole network of contacts influences what people do, not the sight of the smoke, but the whole pattern of instructions that comes into place when you have an emergency.'

The need for authority, and the reaction to the way authority figures behave, is a psychological norm in many fires. 'At Bradford,' says David Canter, 'some people realized it was an emergency when they saw a police officer running because police officers don't normally run, particularly in crowd-control situations. They move steadily as if they're in control. People realized that if a figure of authority was doing something so radically different there must be something serious going on. In emergency situations, if a person in a senior position says, "We need to go this route," then leads that way and starts taking a route that other people wouldn't normally take, then clearly people will understand it's a different situation, the rules have changed and they will follow that, but that senior person has got to get it right. He would kill them if he took them along a route that was closed off.'

Nevertheless the mere possession of a uniform is not enough to create an aura of authority – not in Britain, anyway. At King's Cross, says Paul Godier, head of safety and environmental development there, 'one of the clear lessons was that customer respect for a figure of authority is important. Customers tended to respond to requests from the police but they didn't from members of the London Underground's own staff, even though they were uniformed.'

In fact it could be that mere organization of authority, of people designated to take charge in an emergency who know what they are doing, is sometimes enough to produce order out of potential chaos. At a major explosion at the World Trade Center in New York, says the building's manager Alan Reiss, 'there were people who took charge on every floor and within the groups and made sure that the people who could not walk down the stairs because they were in wheelchairs had someone to assist them, that people who were having asthma attacks had someone to stay with them and make sure that they were OK the whole way down. This was all voluntary, spur-of-the-moment, but I think that you'll find that in these types of crises natural-born leaders appear and people are looking for a leader. It doesn't have to be an executive, it could be the person from the mail room, but someone who knows what they're doing and has a cool head.'

The most agonising results of reliance on authority were seen in a dreadful fire in Our Lady of the Angels School in Chicago in 1958. It affected only one wing of the school, housing 329 out of a total of 1,400 pupils, but 92 students and 3 teaching nuns died, while 76 others were injured in the third worst fire in Chicago history, which was also the third worst school fire in American history. The fire created a myth: it was said that the nuns had imposed a fatal inaction on their pupils who were praying at their desks when they could have escaped.

'It was a very working-class neighbourhood and staunchly conservative and Catholic,' says David Cowen,* 'These were kids whose parents

*Author with John Knowler of *To Sleep with Angels: The story of a fire,* Ivan RD, Chicago, 1996.

basically made their living with their hands, very religious people, people who stayed close to the teachings of the Church. At that time the nuns were very much in control. They were strong disciplinarians and they were attired in their traditional black and white habits and it was an environment in which discipline was the rule. They knew that when Sister said something you did it and you didn't make a scene of yourself, you kept quiet and you didn't talk unless you were spoken to. Much of the directions that were given in those days were reinforced with the subtle or tacit threat of a ruler across the knuckles or a tug on the ear, perhaps a slap on the face or a pull on the hair.' Nevertheless, says Matt Plovanich, then a ten-year-old pupil, 'we were very fond of them. Even if we had a teacher that was unusually rough, we realized that in the nuns' minds this was their family. They did not have a husband or children of their own. We were looked upon as family by them.'

On 1 December 1958, the Monday after the Thanksgiving holiday, 'the fire started,' says David Cowen, 'some time after 2 o'clock it was discovered in close sequence by a number of people, principally two boys, who were returning from emptying waste-baskets in the basement, then the school janitor and then a number of teachers on the second floor. The fire began in the basement at the bottom of a seldom-used back stairway on the north wing of the school,' which was U-shaped with an annex connecting two wings. 'The fire started in a large cardboard drum. It burned undetected, it smouldered, and when the heat from the fire broke a window on the stairway a rush of air sent it new impetus and it spread upwards, caught on the wooden staircase leading to the first and second floors. On the first floor there was a fire door, which prevented the products of combustion from spreading into the first-floor corridor. It travelled upwards to the second floor where there was no fire door but an open stairwell. The inside of the building was entirely of wood, wood lath and plaster – the floors in particular were easy to ignite as they had been varnished with oil-based soaps and waxes through nearly half a century. At the same time the hot air and the products of combustion also spread upwards through an open air shaft that opened at the bottom of the stairway where the fire started and that extended all the way upwards into the attic between the second-floor ceiling and the roof of the school.'

The spread of the fire sparked a secondary fire in the ceiling area above the classrooms, but remained undetected for half an hour. 'By the time the occupants on the second floor finally sensed that something was wrong, it was too late,' continues Cowen. 'The fire and the smoke had begun spreading through the second-floor corridor, the doors of which opened on to this corridor. When the occupants learned of the fire they opened the doors and were met by this wall of thick black smoke. The hallway was impassable at that time so they were forced to stay in the rooms and eventually had to jump out of the windows.'

The alarm was not sounded due a bizarre series of events. Two children and their teacher smelt the fire and went off to find the Sister Superior who alone was allowed to sound the alarm. This delay was fatal. 'The discipline of the school,' says Richard Scheidt, a firefighter who experienced the full horror of the scene, was 'to wait, someone find out, maybe somebody never went to find out.' As it was, says Cowen, the Sister Superior 'was not in her office but substituting for a sick teacher. The teacher giving the alarm first marched her own children out of the building, then ran across to the street to the convent where she lived and called the Fire Department.

'The teachers were waiting for an authority figure to tell them what to do. This was also the case with many of the children.' When the children opened the doors and were met by the smoke, 'the nuns advised them to remain in their seats and to prevent a panic basically had them start saying Hail Marys and told them they were going to wait here for the firefighters who were coming to save them. They couldn't do anything else. Their action was to 'prevent the students from getting up and running out into the hallway which by that time was impassable because of smoke and fire. These women knew that and they decided to have the kids wait there for the Fire Department to come with ladders to rescue them.' Unfortunately that did not calm all the children. Plovanich remembers that 'many of the girls ignored the rosary. They continued to scream for their mothers and fathers.'

In any case, says Cowen, 'within seconds and minutes of learning that the fire was burning, the glass transoms over the doorways of the classrooms broke and the smoke and fire began to spread very rapidly into the classrooms. At that point the nuns abandoned the prayers and directed the students to go to the windows. The sashes were thrown up, they started sticking their heads out the windows and started screaming for rescue and they were fighting for clean air to breathe. I think that the discipline kept them in the room a little longer than might be expected but when the heat started bearing down on them they started to jump out of the windows. It was a twenty-five foot drop and that's quite daunting to an adult, let alone a child. Many of the children were clearly hoping for miracles. 'In the back of our minds,' says Matt Plovanich, 'we all hoped that the cavalry would come and rescue us at the very last moment and it did.'

'From the interviews we conducted,' says Cowen, 'there came a point where panic took over and kids were fighting for space at the windows so that they could get up on the ledge and jump. It's sort of survival of the fittest and the bigger, tougher, stronger kids were able to get up and jump. Meanwhile the nuns, who were very strong, very tough people, did not just sit back and passively watch all this havoc occur. They were up there with the kids hastening them out of the windows. In one case, in one fifth-grade room, I had several survivors tell me how the nun was pushing the kids off the ledges, screaming to them to jump. She was one of the nuns who perished. There were other instances where the kids were very scared and they ran to the nun and draped themselves around her seeking safety and protection and shelter and this particular nun was found later by the fire-fighters to have passed away in the fire and to be surrounded by kids.' Even *in extremis* they 'adhered to the standards that they had known all their lives'. My nun, says Plovanich, 'was going down with the ship in effect, she was leading us spiritually, she did keep the class together. She had the only classroom on that second floor that escaped without a fatality. Even though she was badly burned she blamed herself for the tragedy and had a nervous breakdown.'

Three or four nuns, says Scheidt, 'got all their children out of that building. They lined them up and didn't hesitate, took them right out of that building. There are at least two hundred children alive today because of the action of the teacher.' As for the ones who didn't get out 'you can't second guess what happened'. The nuns 'gave their lives to stay with their children. They might not have had the opportunity to get out as the other ones did.' Cowen and Knowler say simply, 'They did everything humanly possible. By the time that they learnt that the fire was burning it was too late and, as Cowen has to admit, 'There was some reluctance to escape because they had not been given permission to leave. This may have contributed to a delay at least in running to the windows and in jumping.'

# 13
# Get Me Out of Here

> I think that when people die in fires it's not because of panic – it's
> more likely to be the lack of panic.
>               Neil Townsend, divisional officer, London Fire Brigade

As we have seen in most of the cases discussed in this book, one of the greatest problems in escaping from a fire is the sometimes incredibly short period
that elapses between the first sign of fire and the moment when it becomes
an all-devouring demon. The design of some buildings doesn't help. In
Britain and the United States, they used to be designed to give occupants
about three minutes to reach a place of safety. Legend has it that in 1911
the audience at an Edinburgh theatre was evacuated safely in the time it
took the orchestra to play 'God Save the King' twice: two and a half
minutes. And, as we have seen over and over again, exits are too often either
inadequate, because the building has been extended, blocked by furniture
or bric-à-brac, or locked. The latter is a common phenomenon in public
buildings, like hotels and nightclubs, where the management may be
obsessed with keeping out unwelcome intruders. (In Britain this problem
was tackled by the 1971 Fire Precautions Act under which the manager of
such an establishment could be fined £400 or jailed for up to two years for
locking fire exits.)

Then there is the problem of whether in a fire situation you should
encourage people in a high-rise block of flats to stay at home, or evacuate
at the first possible opportunity. There is, as Neil Townsend explains, a difference between practice in the UK and on the other side of the Atlantic.
'In London,' he says, 'we will try and keep people in their homes so long
as they're not directly affected by the fire, whereas the North Americans
will try and get people out to the point that they fit fire alarms in high-rise
residential blocks. It's actually a requirement of the fire codes and they

practise fire drills. Unfortunately there's been the examples of people dying while trying to get away from a fire in a high-rise residential in North America.'

High-rise buildings present their own problems. In a high-rise, the fire can be a long way off yet still fatal: smoke may move up the building as though it were a chimney. The problems can be exacerbated by what is known as the 'stack effect', shown at its most dangerous in a fire in an apartment block in North York, Ontario, in 1995. The fire had started on the fifth floor, and pressure from the gases as they expanded pushed open the fire-resistant door between the fifth-floor corridor and the emergency staircase, allowing smoke to escape into the staircase.

The 'stack effect' is like the updraught that we look for in a good chimney. Air and gases move upwards because they are hotter than the ambient temperature outside. The bigger the difference, the greater the effect. On a cold January morning in Ontario, in a building twenty-nine storeys high, the effect was dramatic – and dramatically increased when a well-meaning resident opened a roof hatch to allow smoke to escape, only to create an even more effective 'chimney effect'.

Another complication is that, for years, fire-safety engineers assumed that people would not endanger themselves by going into smoke-filled areas. Now human-behaviour researchers have shown that we do just that. All the victims to the North York fire were found on upper storeys and in stairways. The door to the burning apartment many storeys below had been left open and smoke had gone up through the floors, via the stairs, lift shafts, heating and air-conditioning ducts. Even though smoke was making the stairs unusable, the investigators found that people continued to try, moving through increasingly severe smoke, to escape by them.

'The problem with trying to evacuate a large residential block,' says Townsend, 'is there are a lot of people involved. If you're looking at a high-rise block with maybe two hundred apartments, if you assume two people in each you've got at least four hundred people to evacuate. The policy we adopt here is that we would only get out the people that are involved directly with the fire, the people whose flat it is and those on either side maybe. [Other] people are perfectly safe. The way buildings are put up nowadays there's no reason why they shouldn't stay exactly where they are. People who try to evacuate down a large building at some stage are likely to become involved in the fire or overcome by smoke. We've had experiences over here where we've tried to evacuate people and we've ended up filling casualty in the local hospital.'

The problems encountered in real life when trying to limit the evacuation of a large building were vividly demonstrated in a recent fire in Boston, witnessed by Professor Ed Galea, fire-safety engineering group, University of Greenwhich. It was a thirty-storey office building housing up to three thousand people. 'A fire started on the twelfth floor,' he says, 'in one of the motor rooms for the ventilation system. Large volumes of smoke were being produced but the fire was contained on this floor and all the smoke was being pumped out of the building.

'The evacuation didn't go according to plan. What was supposed to happen was that the fire floor and the floors above and below should have been evacuated and the alarm system in the building was meant to indicate which floors needed to be evacuated and which floors didn't. The other floors would be warned that there was a fire in the building but they could stay in place until the second-stage alarm system went off, which indicated that they also had to be evacuated. Unfortunately when the people on the higher floors noticed the black smoke billowing out of the building, they forgot the procedures and they all started to try to evacuate. As a result the staircase was congested and it took people some time to get out of the building. In fact some people on the higher floors were queuing up by the fire staircase doors for twenty minutes before they could get on to the staircase. Once they got on to it it was also quite congested and it took many minutes before they started moving down. It's a modern building, with a modern alarm system and, well-trained people but the procedures still didn't work. When the occupants noticed that there was actually a fire and that it wasn't a drill, it didn't quite go according to plan. The building was designed so that people would survive a fire but it wasn't designed so that everyone would evacuate all at once.'

Some people do not behave rationally when faced with a fire – but this is because they were not very rational human beings to start with. John McCool, who was caught in one of the fires started by the gang of youths in Philadelphia described in Chapter 9, describes how 'a lady two doors up, Ann, we went up to get her out because she takes medication that makes her sleep. There was her and her little dog called Cola and she was more concerned with the dog than she was with her own safety. The dog was hiding because of all the noise, so they wound up spending more time in the house

– the back of the house was on fire at the time. I had more concern with them getting out than I was about her damned dog but they did get the dog. Then when we went to get Ann, she said, "I'm going to burn to death." I said, "No, you're not, let's go," and I picked her up and took her out of the house. Then she ran back in and the flames were in her dining room, in the living room, and I had to scream at her. She just froze and then I scooped her up. She'd forgotten her dog was already out.'

But this does not mean that the average human panics in a fire situation. Indeed, one startling fact to emerge from recent studies is that people do not resort to panic when they are faced with a fire. In the past the idea of panic was always linked with the way that people behaved when caught in a fire, and the threat of a crowd becoming dangerously out of control has always been in the back of the minds of regulators and firefighters alike. A useful working definition of panic is 'flight behaviour that is self-injuring or other-injuring and likely to hinder escape'. By contrast, investigators have found that people confronted with fire continued to behave in a calm, rational way until they had no option other than to fight for their lives. Not until choking smoke and flames were upon them did they surge forwards towards the door, blocking the last remaining exit.

'The initial view, say pre-war,' says Neil Townsend, 'was that people panicked to the point that the regulations for theatres and cinemas used to say that any telephone for calling the fire brigade should be sited away from the public so as not to cause panic. There was a great feeling that people involved in a fire would panic and even the media until recently attributed a number of fire deaths to panic. I think that the word panic is used by those who are standing outside and watching the situation, either bystanders or firefighters who are trying to get in. If they can look into the building and see loads and loads of people screaming and shouting and sort of climbing over each other to try and get away from this fire they're going to say those people are panicking.'

Thanks to the work of investigators and specialist psychologists on how we actually behave in the face of a major conflagration this traditional assumption is now perceived as the opposite of the truth. Neil Townsend believes that, 'recent research has highlighted a number of issues, one being the commonly held belief of panic and another is the commonly held belief that as soon as people realize there's a fire they're going to get out and get everyone else out. I think the research is showing that (a) people don't panic.

I think it's lack of panic, if anything, that kills people, that people, and especially the elderly, tend to stay longer than they should within the fire – they may have lived in the house thirty, forty years. One couple I interviewed, the male sat in the front room watching the fire because he wanted, as he put it, to keep an eye on things and that was his way of managing the situation until the brigade turned up. I think if the brigade had taken a minute longer to arrive they would have been looking at another fire statistic.'

'The definition of panic,' says John Keating, 'is usually people doing irrational, impulsive things that may be detrimental to themselves and others around them. What I think is really happening is that people are using the best information they have to try to save themselves and frequently save others. Unfortunately sometimes they don't know the appropriate behaviour, and also their intelligence is clouded by the impact the smoke has on their body chemistry. Look at the autopsies of people who have died in fires by jumping twenty floors out of a window and put the scenario in place: the person sees the fire, sometimes on their own floor, is totally overcome by the smoke and therefore makes their last effort for escape out of a window. People say, "Look, he jumped, why did he jump?" Well, you do the autopsy and frequently these people are almost dead already from the carbon monoxide that they've ingested. It's not that they're making foolish decisions, it's that they're making the best possible decision with limited capacities.'

When the whole executive staff of a major American company was wiped out in a blaze, says Keating, 'they were seen to be clawing their way through a closet door trying to get out of the fire. Well, the fireball came in through the only available exit. The one door that looked like it could be an exit was their last chance of survival, and the first headline said that these executives panicked in a fire. The facts are that there was no alternative, they were doomed. When the fireball came in at the only available exit, they did the only possible thing they could do.'

Victims themselves tend to blame their own behaviour on panic. One problem with a fire, says David Canter, is 'that it's so complicated in the early stages that people will often try to make sense of what is going on without really knowing what is going on in a way that enables them to make a sensible decision. So after the event they look on what they've done and see that many of the things they did were not terribly sensible, and they'll try to explain it by saying that they panicked and that they did irrational, fearful things. People who studied the movement of individuals through buildings, talked to people after the event about what they felt when they formed this view of wild, irrational behaviour, of panic. They were able to show that if you consider the perspective of the person in the situation and look at what sense they were trying to make of it, you can understand that they were acting in what seemed to be the most sensible way in the light of the information available.'

'The first thing,' says Professor John Keating, is that 'it really does a disservice to the people who've been trying to do their best and may have ended up dead. It sounds like these people were less than effective. Second, it leads to false expectations of people who may find themselves in a fire situation and believe that everyone around them is somehow going to hinder their escape and be clawing for their own selfish ends.'

'One of the things that bothers me,' says Professor William Feinberg, a sociologist, 'is that people are using humans, the victims, as the explanation – their behaviour explains why they died rather than that there were not enough exits or that there was not enough warning and a whole host of other problems.'

Feinberg was referring to the tragedy of the fire at the Beverly Hills Supper Club in Kentucky in 1977. It was a Saturday night and the club, a popular dinner and cabaret venue a few miles south of Cincinnati, was crowded with two thousand customers.

It was no ordinary club with no ordinary story. Wayne Dammert, who was maître d'hôtel in one of the dining rooms on that fatal evening, explains: 'It was built in 1937 by a man named Smith. He made a casino out of it and the Mafia wanted that casino and he wouldn't give it to them so eventually one night it burned down. A caretaker and his wife and little granddaughter were staying there that night and they had to jump out of a second-storey window. The two older people were OK but the little girl was killed in that fire. Smith rebuilt it and the Mafia tried to take over again and this time he just gave it to them and started another place.' It became a gambling den until it was closed by government pressure. Under new ownership, it was rebuilt in June 1970. 'It was almost finished, furniture in there and everything, and somebody set a match to it – which was never proved – and it really burned down, a couple of million dollars' worth. Then they rebuilt it again and they opened it and we didn't have a whole lot of trouble with fires or anything.'

To Dammert, it was 'a completely beautiful establishment. Some of the rooms, like the garden rooms, were all in glass, they had big awnings over parts of the room, they had mirrored sections up in the top with flowers growing, they had shrubberies all around the bottom of the glass. The rugs were all thick, beautiful, there were murals on the walls, there were chandeliers throughout the place and the upstairs rooms. We had thirteen or fourteen different party rooms, plus the big show room, and it was absolutely beautiful, draperies, thick panelling on the walls, whatever, it was the best.'

The fire started in a small banqueting room called the Zebra Room, which held about thirty-five people. An electrical fire had been burning behind a wall for some time when it was discovered. It defeated the efforts of staff to put it out and quickly burned through corridors and up the stairs. As smoke and flames spread through the building people moved towards the exits of which there weren't enough. Over the years the original structure had been enlarged so there were no longer enough exits for the number of people the building could hold.

'The fire,' explains Dammert, 'started in the wall, just an electrical fire. They had the wrong kind of wiring, copper wiring against aluminium. Over time it had corroded and caused heat. It did not break the circuit breaker, just set some wood on fire. It finally flashed over and blasted the door open. How it got to the show room [also known as the cabaret room] was unbelievable because it had to come out of that Zebra Room into the main hallway, which was called the hall of mirrors, go up this hall of a couple of hundred feet then make a right-hand turn in through the doors of the show room – like that's where it wanted to go and that's where all the people were.' But not all the people. Dammert managed to alert the many groups in other rooms – among them, a Cincinnati choral group and a meeting of Afghan hound fanciers.

The fire, explains Dick Reisenberg, one of the firefighters on the scene, 'spread through the building basically the same way, through the air-conditioning ducts, through the false ceilings. When the smoke vented into the cabaret room, ignition took place and the fire started. The conditions were right. Now, there was an estimated 1200 to 1300 people, maybe 1400 people, the exact figure will never be known – in that room alone, where 163 of the 165 total victims died so if you're talking figures instead of human bodies an awful lot of people got out, and many more got out than died, but still it's a tragedy.

167

There was a fatal delay in warning the guests after the first signs of smoke. 'People first told some supervisor,' says Professor Norris Johnson, 'and then the supervisor went to check and sent some people to get fire extinguishers.' But 'no one wanted to warn the patrons without first getting authority from some higher level so they began looking for the building owner. As a matter of fact, they went looking for one of the sons of the family that owned the building.' In the Professor's view this was an example of seeking higher authority at the most inappropriate moment. 'In the Beverly Hills there was no one in a position of authority to warn the people in the cabaret room, which is where all save two died.'

The death toll would have been much higher had it not been for an intelligent busboy. 'Thank God,' says Dammert, 'for an eighteen-year-old busboy named Walter Bailey,' who had seen the fire start. 'He saw what was happening, immediately ran into the party room and told those people – it was a barmitzvah party – that they had to leave. He went up the hall to another room and instructed a group of doctors to leave and he went on back to the cabaret room. There were people already waiting in line for the second show, he told those people to go out through the back gardens, he went into the show, he told the lady on the door she'd better go open the fire exits up. She looked at him and then she went and unlocked the fire doors – they were locked because people would sneak in. He stood there and thought, "Man I'm going to lose my job if I tell the people about this, what should I do?" He went up on the stage, took the microphone out of the comedian's hand and said, "Folks, we have a small fire", and he started instructing them how to go and then he gave the mike back to the comedian. In a matter of seconds that fire shot into the room and the people left in there didn't have a chance. Now, if he hadn't gone in that room I'm positive everybody in there, close to a thousand people, would have been killed, so to me he's quite a hero.'

'Most of the people reacted by beginning to move,' says Professor Johnson. 'Most people did not see it at that point as a real emergency because they were finishing their drinks and some of the waitresses were asking them to pay their bills before they left but they generally began to move in the right direction.' Nevertheless 'some people thought that this was simply a part of the act, I think partially because it was the busboy, he interrupted the comedians. Had it been the manager of the building in a dark suit and tie they may have responded in a somewhat different way but

I think you don't expect a busboy in a dirty white coat to appear to give you these directions.' People's reactions, in this and other cases, says Carl Duncan, an independent fire analyst, depend on, one, your level of experience and your perception of the risk, and the circumstance you're in relative to where that fire is occurring. The Beverly Hills Supper Club was a show room. People were waiting to see a show, they were closely packed and in many ways overcrowded but they didn't want to give up their seats for that show. Any public occupancy, assuming that there are activities taking place there, is going to distract people from perceiving how great a risk may be. It's a matter of educating the public to the fact that the fire alarms warn of a risk to themselves, whether or not there are any other physical or visual influences.'

At the Beverly Hills Supper Club, people's reactions to the warning depended on the way in which they had been informed of the fire. In some of the many different rooms in the club, says Professor Johnson, 'someone walked in, said, "There's a fire, please leave", and people walked out quickly. In one area someone said first, "There's a fire, you should leave", then someone else came in and contradicted that because they didn't want to create panic. Moments later someone else said, "You must leave", and people were confused. In those areas where there was a direct, immediate warning, people left easily and generally had no problems. In certain rooms in the front, in the bar area, they saw the smoke and they left on the basis of that physical evidence rather than being warned.'

The headlines next morning read 'Panic kills 300'. In fact the death toll was 'only' 160. But the questions remained: why was the death toll so heavy? Was it because people had panicked? The sociologist Professor Norris Johnson examined the evidence, interviewed witnesses and tramped round the ruins of the club. 'What most people perceive and what most of the popular reports have said is that mass panic ensued. People were concerned about no one other than themselves. We found no real evidence of that. If you define panic in terms of non-rational, individualistic, non-social behaviour you certainly couldn't argue that panic occurred in this situation at the Beverly Hills and most research suggests that it doesn't occur in similar situations. Certainly people scream, shout and are frightened but even when they're frightened people tend to continue to behave in social ways.

'Clearly, near the end people were competing to try to get out first but they were still helping one another. There's one dramatic account of this older woman who fell. She tried to get people to get someone under her out before they rescued her. Throughout there were altruistic responses like this. Even those who were severely injured reported being assisted, assisting others.' The biggest surprise was 'the extent to which strangers were aided and people were themselves helped by strangers. People referred to "this man that I didn't know who encouraged me and who tried to help me out".'

When he and his colleague William Feinberg analysed the fires, they looked at all the possible relationships between those who died and the survivors. They felt that, in a panic, says Feinberg, 'we would expect to find that employees had left immediately and would survive much better than the patrons because the patrons had to negotiate some difficult exits. We did not find any difference in those rates of dying. We expected to find that women would be dying at a higher rate than men because physically men are better able to compete. We did not find any difference between men and women. We looked at those couples in which one survives and one dies. In lots of cases both husband and wife died but in those in which one survived, 70 per cent of the dead were husbands. Clearly the men were taking into account their responsibilities as males, as husbands and fulfilling those roles and making sure that their wives got out. A 70:30 breakdown like that, it's a simple statistic but a dramatic one in that it tells us that the social order was strong and people were committed to carrying out their roles.'

If a single overriding impression emerges from virtually all the accounts of fires it is that people, in general, behave not only rationally but decently and often unselfishly. At the explosion at the World Trade Center in New York, where people had to wait up to three hours to be evacuated, office colleagues helped each other through the smoke, sharing scarves to cover their faces and counting then shouting out the number of stairs in each flight to help those behind them. Nelson Chanfrau, in charge of emergencies for the New York Port Authority, says, 'People appeared to be fairly calm. In a building such as this you have nearly 40,000 occupants as well as several thousand visitors and shoppers going through the concourse. In a situation such as that you would expect to see people not behaving as well as they were during the evacuation. Everything was done almost in a militaristic way. We heard stories as we interviewed people after the bombing of people being carried down in wheelchairs, people who were not feeling well, pregnant women and at the head of the stairs someone would yell out, "Coming through with a pregnant woman", and people would just go to both sides of the stairways and allow the person to be brought down in a wheelchair.' There was 'a tremendous amount of co-operation and camaraderie, if you will, something that, perhaps unjustly, New Yorkers are not known for. I personally can't remember any incident that was relayed to us afterwards about anyone running out and knocking people down or not going along with everyone else.'

Similarly, when investigators studied the film of the Bradford football stadium disaster, they found, says David Canter, that 'People helped each other. You even see in the film that people bump into each other, then sort of apologize whilst they're trying to get out. The normal conventions of daily behaviour do carry through in those emergencies. We can see that clearly in the Bradford fire. We can see clearly too that some people fell over and in the confusion of trying to get out others stepped on them, but this wasn't deliberate. After the event they may feel terribly guilty about stepping on people who'd fallen over.'

At the disaster at the MGM Grand in Las Vegas, says Tom Klem, a fire protection engineer, 'people were in various stages of activity, some were sleeping, having gambled all night and others were awakened some time during the fire by sirens, Fire Department horns, sirens on the exterior of the building. There was a complete mix of what people were doing, where they were and so forth. But one common thing that we found in the investigative process is that people joined one another, they had a tendency to exchange information and plot routes and strategies for what they could do to intervene in the products of combustion that they saw moving towards them. There was some clustering of people within rooms, exchange of information.'

A survey was conducted of the survivors of the fire and found, says John Keating, emeritus professor of sociology, 'that people who knew how to behave in a fire, for instance, wet towels and put them under the doorway, made sure that their bathtubs were filled with water and so forth, would frequently reach out in the hallway – these were the smoke-compacted upper floors of the hallway – and pull people in the room, shut the door behind them and refortify the room. We found over sixteen or seventeen or these clusters of safety, some of whom had ten people waiting out the fire. The frequent response of the people that were pulled out of the hallways was that if someone hadn't come out and pulled them into a room and then fortified the room, they would be dead now because they would have inhaled too much smoke. What this emphasizes is that if people know how to behave they'll behave appropriately during an intense situation, and at the same time they'll be attempting to be as altruistic as they can to other people who may not be as fortunate with the knowledge that they had.' Indeed, some people stayed in their room with strangers for as long as three hours before they were rescued.

For me it was the World Trade Center explosion that provided the best example of human behaviour at its finest. Kathleen Collins is a wheelchair-bound lawyer: it took six hours for her to be evacuated from her sixty-sixth floor office to the pavement. 'I was concerned,' she says, 'but I have a brother who's a captain in the Police Department, a brother who is a fed-eral agent and I have my brother's brother-in-law who was a firefighter at that time down in Lower Manhattan. So I kept saying, "Well, one of them's going to come and get me, so I figured somebody's coming up so, you know, just take your time and don't panic. Your worst enemy is yourself, so just stay calm." Everybody else was staying calm so that made us all feel better, which helped.'

# Postscript: A Model Future

Engineers and architects can design safer buildings. Instead of assuming that the people who inhabit a building will behave like ball bearings in a mechanical model, they can take account of what people really do and design buildings according.

Neil Townsend, divisional officer, London Fire Brigade

In the past, it was assumed that people would panic at the first sign of fire, so architects, builders and regulators went in for what Neil Townsend calls 'a very prescriptive building with lots of fire exits and without any real understanding of how people behave. Once they ditch this sort of mythical term of panic, then we can start constructing more effective and more cost-effective buildings that people can get out of in the event of a fire. That includes people's homes. If you study how people behave in these different places, then you can build in the appropriate fire-safety arrangements – the right type of fire alarm, and number of exits. You study how far people must travel before they're out of the building. Just by studying their behaviour you're adding to the argument of building designers that we should construct our buildings around people and not just fit people into buildings. In the past we've said, "There will be a fire exit at every so many metres", depending on what type of building it is, and often it's not required. I think at times we're a bit onerous on building developers.

'In the United States regulations for new buildings now specify that the main entry and exit – the one used in normal situations – is large enough to cope with the emergency evacuation of most of the building's occupants. In the past, as one expert put it, 'engineers designed fire-safety systems that ignored the people inside the buildings and then when the systems failed, as they inevitably did sometimes, the people were expected to behave like superhumans to save themselves.'

The earlier naïvety was astonishing. David Canter says, 'I first became interested in what people do in a building on fire when I was working as an architectural psychologist, looking at the way the design of buildings could help people to use those buildings more effectively. I remember talking to an architect about a very interesting building design and he said to me it would never be allowed because of the fire regulations, and I was very surprised because it made a lot of psychological sense in terms of a building that people could find their way around easily. When I looked into the background I found that all the information that was being drawn upon to shape buildings was based on assumptions about what people would do if the building caught fire. A tremendous amount of money is spent on limiting the shapes of buildings, putting in fire doors, limiting the lengths of corridors – a whole variety of things shape our surroundings because of assumptions about what people will do if that building were to catch fire.

'When I examined this material and looked at the basis of these guidelines I found that they had been drawn from a very few incidents in which fire officers and engineers had looked at what people were doing at the time that the fire engine arrived, when the building was in flames. There had been no systematic research of people's actions right through the development of the fire. These few incidents had been used to shape all the guidelines and had a big influence on the whole shape of our cities. I decided it was important to try to get detailed information of what people *did*, to challenge these assumptions that were floating around that were stopping us getting all this important information into the design and construction of buildings.' (Though even Canter has to admit that the regulations are effective because few people die in fires in public buildings in the UK.)

What is now being envisaged by an increasing number of scientists and investigators is a major shift from buildings where designers have simply to conform to regulations to those where the designers think of the way in which people behave. As B. J. Meacham puts it,* 'Many fire engineers consider building occupants as "items" and not as people.' They ignore the different factors that matter before and during escape from a fire. They 'ignore many important human responses, physiological and behavioural

*'Integrating human-factor issues into engineered fire safety design', from *Human Behaviour in Fire* – proceedings of the First International Symposium, 7 September 1998.

174

factors as part of their fire-hazard assessments.' They assume that as soon as the alarm sounds all the occupants will begin evacuation, that they will 'walk at the same speed, and all occupants will reach a place of safety within the calculated time'.

This, as we have seen time and again, is an unrealistic scenario. Will occupants see or hear a warning (especially in a noisy building)? Will they recognize the fire alarm, and if so will they take it as real and not as a false alarm? (Researchers have proved time and again that people don't respond well, or at all, to non-voice signals). Then there's the – often dangerously slow – reaction. As we have seen, people have to 'validate' a fire, seeking advice; they have to define the danger then decide how to behave – stay or leave.

'Before psychologists started studying what people actually did in a fire,' says David Canter, 'the thinking was very mechanical, almost as if people were a pile of potatoes shuffling down a chute and would just perform in a mechanical way without any thought or understanding of the relationships between them. All the planning and decision-making was based upon the calculations of simple flows of people through buildings, working out the spaces and dividing it by the width of the average person and just seeing it as some totally unthinking flow problem. Of course, what psychologists have done is to draw attention to the fact that human beings, even in the most difficult and dangerous circumstances try to make sense of their surroundings.'

There has been little research into the relative effectiveness of different warning signals. It emerges that bells are far less effective than verbal warnings. 'People would listen to words as opposed to simply following arrows or direction signs,' says John Keating, a professor of social psychology 'and, in fact, arrows and directions are notoriously lacking in any building because they put the exit sign above the doorway and the first thing to be obliterated by smoke is the exit sign. They'll tell you not to use the elevators but give you no clue about where the stairs may be located, even though you had entered and exited the building for the last twenty years by the elevators and have no awareness of where the stairs would be, and, of course, visitors in a building are totally unfamiliar with the layout.'

To make matters worse, the American legal system ensures that hoteliers in particular feel, as John Keating says, that 'the instructions must cover all the potential contingencies. Many hotels, at least in the States, feel they'll be liable if they omit one contingency so you have very elaborate fine-printed instructions on how to behave in a hotel that I suspect none of us reads or they'll have ridiculous types of things like bring your bungee cord to escape out the window and tie it on a radiator. I've seen those written for travellers, saying, if you go to a hotel make sure you take some evacuation modes, like a rope-ladder to climb out the window. People won't do that. I think there should be four or five clear bold-print items on the back of every door telling them to locate the nearest exit, take your key when you go – the essentials, as opposed to every possible contingency that can happen. One time the fire marshal of the State of California and I figured out about six key words we should put in every hotel room.'

Even the best traditional alarm systems do little or nothing to motivate people to evacuate a building. Research has proved that systems that provide information about the size and location of a fire are better at motivating people to leave. Some modern systems can give such information, either through visual display units or audibly by recorded (or computer-synthesized) speech. The use of 'mimic' displays, in which the position of a fire is shown clearly in relation to a plan of the building concerned, can speed up the interpretation of the complex layouts found in so many large structures.

The great advances in the science and engineering of fire protection have already shown some encouraging results, both in terms of hardware and of training. 'There's a big move towards smoke detection,' says Neil Townsend. 'It's something that fire brigades up and down the UK have been pushing for years now. It's been included in the fire codes that new homes will have smoke detectors fitted and in London especially we're trying to move towards having domestic sprinklers installed. I think that the smoke detectors will alert people and the sprinklers would control the fire. Those two measures combined would significantly reduce fire deaths because over 90 per cent of all people who die in fires die in their own homes.'

And, of course, King's Cross had a marked effect on training the staff of London Underground to cope with fires. At the time of the fire, says Paul Godier, head of safety and environmental development, London Transport, 'different staff had different roles and just as a customer wants to carry on doing what it was they originally intended to do, going home or whatever, so a member of staff who has a particular function will try to carry on with that function. Ticket collectors tried to carry on collecting tickets. What we've introduced since then is a multi-role cadre of staff who may be on the gate line for part of their work but may spend time selling tickets or on platforms, patrolling the station or whatever. Given the training they've had in their role in an emergency plan it now doesn't seem to them like a deviation from their normal role if they begin to participate in an emergency evacuation, it seems like part of their normal job – indeed, it *is* part of their normal job. You don't get this inertia that was evident on the night of King's Cross where people tried to carry on with what they were normally doing.'

After staff training comes an attempt to improve the behaviour of the passengers (now, for managerial reasons, known as customers). It's very important, says Paul Godier, 'to ensure that customers know what to do. Nothing is worse than customers being confused about what to do. There are two things that help: a universal public address system within stations and on trains means that an immediate instruction can be given to customers; and greater use of signs, including emergency exit signs, within the stations, means that customers have some information immediately available to make the right decisions about what to do and then get on and do it.'

But perhaps the most sustained and successful attack on traditional alarm systems came from John Keating. In 1974 he and Elizabeth Loftus devised the first vocal alarm evacuation system for a high-rise federal building in Seattle. It has now been widely copied. It 'had an alarm that was a very well-heard alarm. It wasn't a bell, it was a tone and we had a woman introduce the message by saying, "May I have your attention, please", twice. We used the woman to be a calming effect but then we switched to a male voice for two reasons – one, if one voice switches to another people pay more attention to it, and, two, the stereotype is that the male would be the one directing people during a disaster or fire. Then we repeated everything twice but in different words: if you hear a message and think it's being repeated you tune out. We used words that were most frequently used in the English vocabulary because the words that we hear a lot and read a lot are the ones we understand. We tried to do this with a very calm voice. We found it very successful.'

They didn't use a bell because, he says, 'If you hear a bell it means there is going to be a fire drill or that there's something that you don't have to worry about like the end of a class. We believe a bell would take people a long time to respond to because it is most typically related to a false alarm.

One thing I've been trying to say repeatedly is that you should get the ambiguity out of the situation and the bell is ambiguous at best.' So the signal they used, the 'federal communication systems alert signal' is 'a distinctive sound that's like a mini-horn that comes over our radio channels to say, "This is an emergency test." People hear that, they don't know quite what that means but they know that it's an emergency.'

But they couldn't get away from the dread word fire, however much they tried. 'Historically,' says John Keating, 'there have been events where people would burst into a crowded environment, yell, "Fire!" and people would jump out of their seats, not know what to do, run at doors then pile up and eventually not escape. Many fire jurisdictions feel that you shouldn't use the word fire. We tried a variety of other words experimentally with students but the word fire was the one that triggered unambiguous awareness of what the situation was.'

All the researchers are aware that things have changed for the better over the past few decades, and that today a new approach is required. 'What we're really looking for is the next stage,' says David Canter. 'We've dealt with the great majority of problems caused by problems in design and obvious dangers that can be created by people not managing or organizing their buildings effectively. What we have now is a residue of problems caused by people not understanding human activities in fires. We've sorted out most of the hardware problems, it's now the software problems, the "liveware" problems, what some people call underwear problems; those aspects of the building that underlie its use are the things that need to be dealt with. We have to shape the building to take account of what it is that the persons themselves will do rather than just think about the building as a mechanical device that is lived in by uniformed, unintelligent pieces of mechanism.'

As a result, he says, 'The research is moving on from specifying where doors ought to be and how thick windows ought to be and those physical aspects of design. It is forcing us to move on to management issues about how a building is organized, on to training issues to look at how people learn about what goes on in a building and how they can use that in relation to emergencies, to instruct people in what they should do in particular types of emergency situations and how they can act as a consequence of that. It's interesting that it's actually fed into the whole management of safety in dangerous environments. A lot of the places we've been talking about – football grounds, hotels, hospitals – these are places that shouldn't be dangerous at all. There's nothing particularly dangerous going on there. A big steel mill or a petrochemical plant is potentially a very dangerous place, though, and a lot of what we've learnt about how people deal with emergencies in these day-to-day situations we have now transferred to help in the management of safety in industry.'

Perhaps the most promising line of research is the expanding ability to simulate fires and thus provide far more realistic – and universal – training for investigators than had previously been possible. Computer science is providing opportunities that in the past could only have been achieved by practical experience at the scene of fires. The ATF, for instance, has developed a virtual reality fire scene on a CD-ROM so that they can have 'practical' experience without leaving their computer stations. As Mike Brouchard, chief, arson and explosives program, ATF, explains, 'This training tool will bring the fire scene to the investigator's computer. It'll enable the investigator to walk through a fire scene systematically as we expect they should in real life. You can walk through a fire just like you're arriving at a fire scene, scan the crowd, pick out witnesses who you think may be important, look at the exterior of the scene, move throughout the fire scene. You can back up, assess the scene from a distance, look for the least amount of damage and the most.

'The investigator can then go systematically through each possible source of ignition and if they're stuck where they have a question on electrical fires they can go to a tutorial section where they'll have an expert on electrical fires talking to them on a tape sequence in this computer, or they'll have a library they can refer to with articles of interest that can help them identify possible causes of this fire. This will replace what we can only teach in class-rooms.' Perhaps most impressively, 'We expect about 80,000 people will be trained with this tool this year.

'In the past investigators would look at videos or still photographs and lately digital photographs, which allows them to come into a room, look 360 degrees around just like with their own eyes, assess different aspects in the scene. There are no distortions, everything's true to life, it immerses you right inside the scene.'

Over the past few years computer modelling has emerged as possibly the most promising of all the new tools in research into behaviour during fires. As a result, in theory at least, the buildings of the future will be constructed using computer modelling. One of the most advanced research projects is at the University of Greenwich, where, as Professor Ed Galea points out, their model includes the way people behave: 'We're trying to take into account fire response as well as just ordinary behaviour of people during an evacuation. In that respect we're unique in that in the exodus modelling environment we've developed we can incorporate fire, smoke, heat and toxic gases.'

Before the development of the present generation of computer models, he continues, 'it was really a case of running a full-scale evacuation drill to see how a building would perform. However, in that case you've got the problem of not running a realistic scenario because you can't expose people to heat, smoke and gases and so on. Otherwise people did simplistic calculations, so-called flow calculations. They said, "Well, this many people will flow out of an exit of this size in a certain amount of time", and they took these flow calculations as being representative of how the building and the population would perform. However, people don't behave as fluids or as ball-bearings, as these models assumed, so they were unreliable in terms of their realistic reproduction of what happens. The primary problem is that they excluded human behaviour from their accounts.'

'People often think that attempting to model human behaviour in the event of evacuations is impossible,' he says, because it was thought that people panicked and 'behaved in an unpredictable way. That's not true. If people did panic and act irrationally then it would be virtually impossible to simulate that reliably but we now know that people do not panic and behave irrationally during evacuations, which makes it possible to model the situation. We look at what people actually do under real fire conditions and we take that behaviour and we try to build that into our computer simulations.

'People behave in pretty much the same sort of way during most evacuations. At the sound of an alarm system they take a certain amount of time to comprehend what the situation is, they then decide what to do and then they move to the evacuation. The initial response of an individual is quite important to include a simulation model and we need to be able to delay the response of each individual in a simulation by an appropriate amount before we set them off. Then it depends on the nature of the building that the people are in, whether the people are experienced and trained in evacuation procedures or whether they are naïve about the procedures – for example, whether you're simulating an evacuation from an airport terminal where people would be naïve concerning the evacuation procedures as compared to the procedure in an office building where there are regular drills and people are regularly trained in how to respond to an evacuation. You need to take these factors into account when you're attempting to simulate the evacuation process.

'We need to feel reasonably confident that what we're simulating is actually what happens in the majority of situations so we try to compare model predictions with experimental data wherever possible. We attempt to reproduce certain drill situations where we know the precise situation – we know how many people were in the building for example, we know which exits they used, we know how long it took them to evacuate. Then we set up that case in the computer model and we try to reproduce it. Now we can take it a step further in the computer model. Obviously in drills you can't expose people to real hazards but in computer simulations we can. We have data on how toxic gases and heat have affected laboratory animals, and we can extrapolate that to human response, then build it into the computer model. We can run evacuation simulations of people in buildings that are exposed to hazardous situations.

'An example of that would be how people behave in smoke. We have some quite good data on this from Japan where a series of experiments was conducted in which people were exposed to differing smoke concentrations. The Japanese didn't just use theatrical smoke, they used irritant smoke so that these people's eyes hurt and it was making it difficult for them to breathe. Then their behaviour was measured and quantified. That information is now built into the computer model. By comparing our model against predictions and building into it data that we've found from experimentation we can simulate realistic situations.'

Galea's model can also be used on existing specialized buildings like airport terminals. When the majority of people go on holiday, he says, 'they're more or less brain dead. When they arrive at the airport they don't know where they are. They're trying to find the check-in facility and their mind is on other things. Did they turn the gas off? Did they turn the electricity off? Did they tell the milkman not to deliver any milk? They're not really thinking about the situation that they're in. The buildings are very large and complex, there are lots of people around, lots of activity going on. It's a very confusing environment, with lots of signs and it's difficult to identify the emergency signs. When you're in one of these situations it is difficult to evacuate from it. We would need to set up those conditions in the computer model to reflect what's happening in that situation. For instance, in those situations people mightn't use the nearest serviceable exit, they're more likely to use the exit they're most familiar with, the one they came through into the building. There may be procedures in place to enable occupants to find the nearest exit – maybe staff will direct people out and so on – but all of these things need to be taken into account. Now, an office building is a different situation. In one office building people might be very familiar with the structure, they use it every day, and they are trained in how to respond to the alarm system, they know where the emergency exits are and so on. In that case you would set the model up to reflect that situation, so you could direct people then in the computer model to the nearest exit rather than the exit that they usually enter the building through.'

The newer models can even take social groups into account and, says Galea, 'the context of why they're in the structure. People don't behave like ball-bearings, they don't evacuate independently, especially if there are social links between the individuals involved. A family group will try to evacuate as a unit and the senior member of the family may shepherd the others out of the structure. You may even have situations where a family group is separated in a structure during an evacuation and the senior member will go in search of the missing ones rather than evacuating. You could get quite complex flow behaviour happening, contra-flow situations where people are leaving the structure while others are entering it, trying to find the missing members of their group.

'Applying computer models to the design of buildings is the main focus of the work we're doing. We want to help design the building so that when it's a bright idea in an architect's mind or on a computer screen, we want to put in the computer model then look at how the structure is going to respond, how the people are going to respond to it in an evacuation situation. We'll hopefully be able to lay out the structure, develop the procedures, perhaps move the exits into different locations so that we make evacuation very efficient.' But in the hands of amateurs, though, computer models can be dangerous. If they aren't applied correctly the results they give could end up set in concrete, which could prove fatal.

'Evacuation is not simply about the structure, or about the configuration of the building, how wide the exits are, what's the maximum travel distance. Evacuation concerns four complex interacting components: it's the configuration of the building, the environment – are people evacuating through smoke, heat, toxic gases, debris – it's the procedures in place and whether people are trained in how to evacuate the structure, and, most importantly, it's about human behaviour and how people respond to the changing situation.

'In the past we've relied on only one of these four components: the configuration of the building. We need to bring them all into account to understand evacuation and they need to be taken into account in our computer models that try to simulate evacuation. If you don't take these components into account then you're not modelling what happens in reality.

'As buildings become more complex in design and usage, it's becoming of paramount importance that modelling technology is used in their design, primarily because we don't have experience of these buildings, how they're going to perform, so we can't rely on the old regulations and rules of thumb that we've applied in the past. As buildings become more innovative in their design we're going to need to rely on modelling technology to try to predict how the building and the population are going to perform. I'm thinking of large, complex bases – for example, the Millennium Dome in Greenwich, London. It's a large space, it contains buildings within buildings and the circulation space is complicated. Airport terminals are complicated, with shopping malls built into them. Using the old methods, it is difficult to judge how these structures are going to perform, whereas computer-modelling technology will enable us to test lots of different types of scenarios that are likely to occur in these structures and hopefully to design out the problems that we may face.'

Yes, fires will always break out but we now know how people and buildings can be better protected. Human behaviour in fire situations is not unpredictable; it is, however, highly adaptive, varying from one scenario to another, but fits into recognized patterns that can be measured and accounted for. All of the patterns of behaviour we have seen can be incorporated either into fire-safety training, or into the design and engineering of individual buildings.

People can be given longer to escape – and scientists are investigating ways of getting out so that they have more time; some new buildings have been constructed with this in mind. Fire drills can train people to use fire exits: if they trust them they will use them in an emergency, instead of blindly trying to get to a familiar route. It's no good putting in the required number of fire escapes if they are never used: we have to design safety features in buildings according to the people who will use them, not expect people to fit into the building's design.

We can also train people in a way that utilizes the strict social hierarchies and roles we play in society to help us escape fire. In an office, if the managers take charge and organize in an emergency they will probably be listened to. In restaurants, the staff should be encouraged to take the initiative in evacuations.

Perhaps the most important thing to remember is that when people are caught in a fire they do not radically change the way they behave. This means that if a safety system requires of its user a significantly different pattern of behaviour it is unlikely to work. Instead, it is incumbent on architects and builders to ensure that existing structures and escape routes conform to normal human behaviour. But however successful and ingenious the advances in fire protection, the object of investigation remains the same: to ensure, as Mike Brouchard says, that a fire disaster is 'something we can resolve and hope that it won't happen again'.

# Index